CHEMICAL ENGINEERING METHODS AND TECHNOLOGY

AQUEOUS ELECTROLYSIS UNDER EXTREME CURRENT DENSITIES

APPLICATION TO MICRO- AND NANO-SYSTEMS FABRICATION

CHEMICAL ENGINEERING METHODS AND TECHNOLOGY

Additional books in this series can be found on Nova's website under the Series tab.

Additional E-books in this series can be found on Nova's website under the E-book tab.

NANOTECHNOLOGY SCIENCE AND TECHNOLOGY

Additional books in this series can be found on Nova's website under the Series tab.

Additional E-books in this series can be found on Nova's website under the E-book tab.

CHEMICAL ENGINEERING METHODS AND TECHNOLOGY

AQUEOUS ELECTROLYSIS UNDER EXTREME CURRENT DENSITIES

APPLICATION TO MICRO- AND NANO-SYSTEMS FABRICATION

ROLF WÜTHRICH

AND

ANIS ALLAGUI

Nova Science Publishers, Inc.

New York

For permission to use material from this book please contact us:
Telephone 631-231-7269; Fax 631-231-8175
Web Site: http://www.novapublishers.com

Library of Congress Cataloging-in-Publication Data

Wuthrich, Rolf.
Aqueous electrolysis under extreme current densities : application to micro and nano-systems fabrication / authors, Rolf Wuthrich and Anis Allagui.
p. cm.
Includes index.
ISBN 978-1-61122-038-4 (softcover)
1. Micromachining. 2. Electrochemical cutting. 3. Density currents. I. Allagui, Anis. II. Title.
TJ1191.5.W825 2010
671.3'5--dc22

2010035919

Published by Nova Science Publishers, Inc. † New York

Contents

Chapter 1

Introduction

Electrolysis has been known since the early seventeenth century. Today, it plays an important role in a large number of industrial processes. The possibility to control the reaction rate with electrical parameters is one of the aspects making this technology so interesting for many applications. Electrolysis is seen in its traditional forms, as well as high pressure or high temperature electrolysis. A third variant, the object of this book, is the electrolysis under very high current densities, typically of $1\ A \cdot mm^{-2}$. A wide range of applications are known such as wastewater treatment [1], surface engineering [2], micro-machining [3, 4], and nano-structuring of materials [5]. One notable new effect occurring under such extreme conditions are so called electrochemical discharges, which are electrical discharges occurring through a gas film formed around the working electrode. Several names are used in literature for this phenomenon, such as anode or cathode effects, depending on which electrode the discharges happen, non-equilibrium plasma electrolysis or contact glow discharge electrolysis.

The history of electrochemical discharges can be considered as a succession of rediscoveries of the same phenomenon by different scientific communities. In the 19th century the effect was studied by physicists as a light source with a rich spectrum. Not much later, in the beginning of the 20th century, electrical engineers exploited it for the development of non-linear electrical components. Later on, the chemical and electrochemical community used the electrochemical discharges as a source of non-Faradaic electrochemical reactions. In the second half of last century, micro-precision machining of non-conductive materials, surface treatment of metals and recently, the fabrication of ultra-fine particles has been reported as promising emerging applications of the electrochemical

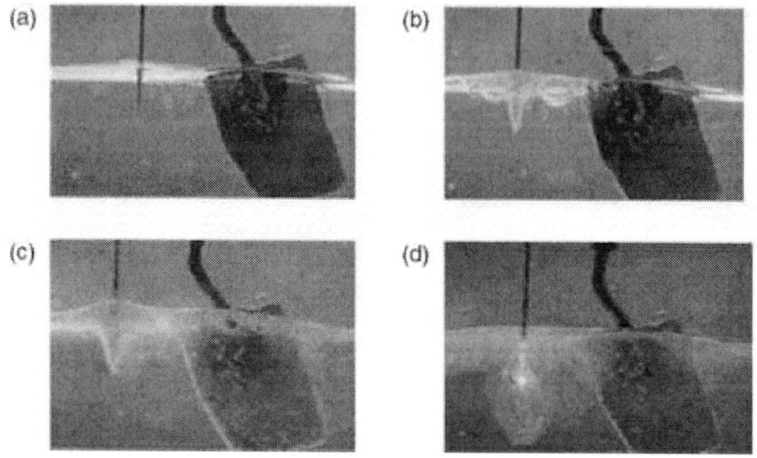

Figure 1.1. Successive steps towards the electrochemical discharge phenomena: (a) 0 V; (b) 7.5 V; (c) 15 V; (d) 35 V. Two electrodes are dipped into an electrolyte. The terminal voltage is progressively increased from 0 to 40 V. At around 25 V a gas film is formed around the cathode, and at around 35 V the electrochemical discharges are clearly visible. Reprinted from [4] with permission from Elsevier.

discharge phenomenon.

For a first approach to the phenomenon, consider an electrochemical cell as illustrated by figure 1.1. The chosen configuration is such that the working electrode has a much lower geometrical surface than the counter electrode i.e. the current densities on the counter electrode will always remain well below 1 A/cm^2. For low cell terminal voltages, traditional electrolysis occurs. As the voltage is increased, the current density also increases and more and more bubbles are formed. In chapter 2, it will be shown that the growing number of bubbles and their mean radius increase with current density. When the terminal voltage is higher than a critical value, the so-critical voltage, the bubbles coalesce into a gas film. Light emission can be observed in the film where electrical discharges occur between the electrode and the surrounding electrolyte.

The following relates selected highlights of the exiting history of the phenomenon. A more detailed review about the history of the early applications of electrochemical discharges can be found in [3, 6]. The first report about electrochemical discharge phenomenon is generally attributed to the two French physicists, Fizeau and Foucault [7]. In 1844, they described in their paper entitled "Recherches sur l'intensité de la lumière émise par le charbon dans l'expérience de Davy" (Study of the intensity of the light emitted by the carbon in the Davy experiment), the observation of light emission in an electrolytic solution emanating from a cathode during electrolytic water decomposition by means of two electrodes of platinum supplied by a constant potential difference. Subsequently, many researchers have been inspired by their discovery as a potential source of light spectra and heat generating electrochemical system.

In 1893, Hoho [8] with the collaboration of Lagrange, reviewed the phenomenon more deeply and described the intermittence of the discharge activity and light emission, as well as the unstable region in the electrical characteristic. They reported an interesting application of the high calorific source property of the discharge phenomenon. The experiment consisted of polarizing catholically a stack of alloy steel which was subject to heating process depending on the gradually increased cell terminal voltage. They observed a change in the molecular structure of the electrode, but only at the surface [8]. Recall that conventional hardening elevates the temperature of a material both at the surface and in the volume. Consequently, when contacted with a cold liquid the whole material will be hardened. Various applications followed the publication of their paper, especially in the metallurgy domain.

One notable technical invention using electrochemical discharges in the field of applied physics is the current interrupter apparatus presented in 1898-1899 by Wehnelt [9]. The German physicist had proposed the first generation of electrolytic interrupters. Wehnelt's apparatus consisted of a large lead plate, placed in dilute sulphuric acid, as a cathode and a platinum wire tightly fitted into a glass or porcelain tube as an anode. Such an arrangement, when inserted in the circuit of a primary coil, gave rise to a rapid intermittency in the primary current. The frequency of the Wehnelt break can be adjusted by regulating the extent that the platinum wire protrudes through the glass or porcelain tube and the inductance of the coil. Wehnelt current interrupter had many applications ranging from X-ray imaging to wireless telegraphy.

In the middle of the 20th century the same phenomenon was used by the chemical community as a source of non-Faradic electrochemical reactions. The British authors Hickling and Ingram [10] introduced what they called contact glow discharge electrolysis (CGDE) of aqueous solutions. This phenomenon was observed when they were studying the electrolysis of liquid solutions of ammonia while gradually raising the applied voltage between a wire of platinum and a counter electrode of much larger surface. A remarkable feature of CGDE is that its chemical effects are similar to the ones observed with α-radiolysis of water, in particular their yield is several times the value of Faraday's law. The chemical products are novel for conventional electrolysis such as the release of large amounts of H_2 at the anode, O_2 at the cathode and H_2O_2 in the liquid phase, among other secondary reactions output.

In the second half of last century, electrochemical discharges have attracted the interest of engineers for micro-scale precision machining of non electro-conductive materials especially for glass and composite materials. The process is known, among other terminologies, as Spark Assisted Chemical Engraving (SACE) [3, 4]. SACE micro-machining uses the heat source property of the electrochemical discharges to promote local chemical etching of the work-piece. Here, the heating process joins the technological application previously introduced by the experiments of Hoho for metal heat treatment. Promising applications are foreseen for glass micro-machining for the benefit of MEMS (Micro-Electro-Mechanical Systems) with applications such as, micro-reactors, micro-mixers, micro-pumps, lab-on-chip or drug delivery devices.

Another fundamental application in the field of synthetic chemistry has been reported by Harada et al. [11–13] in the last quarter of the 20th century, where organic compounds have been implemented in the experimental conditions of electrochemical discharges. Harada et al. discovered new routes for the synthesis of amino acids. A carbon rod, as the cathodic polarized electrode in an electrochemical cell, in aqueous solution of NH_3 and under an electric current of a few mA during short time periods, lead to the formation of amino acids and urea. The amino acid analysis plot [12] shows the detection of glycine in the first place with other amino acids such as aspartic acid, threonine, serine, glutamic acid, alanine as well as urea. In another report [13], they conducted series of experiments with solutions of aliphatic amines and formic acids and they showed as well the formation of some of the essential amino acids such as glycine, alanine, aspartic acid and glutamic acid.

The most recent application of the electrochemical discharges is the synthesis of nano-sized particles. First attempt to use this novel technique in aqueous solutions has been introduced by Wüthrich et al. in 2006 [14]. The authors demonstrated a facile technique to produce platinum and gold nanoparticles using this phenomenon.

The following chapters will present the fundamentals of plasma electrolysis, discussing the physicochemical effects responsible for their onset as well the chemical and electrochemical reactions taking place during the phenomenon. In the second part an overview of the current applications with particular focus on micro- and nano-system production using this effect is presented.

Chapter 2

Gas Evolving Electrodes

The bubbles formed at a gas evolving electrode affect the current-voltage characteristics of an electrochemical cell in various ways. Modeling these processes remains challenging, and mainly only qualitative models are available. The formation of bubbles highly depend on the considered electrochemical reaction and for shake of simplicity, the discussion will be restricted for the case of bubble formation by water electrolysis. For a more detailed discussion of gas evolving electrodes, the reader is redirected to more specialized texts such as [15].

2.1. Current-Voltage Characteristics

Consider a two electrode set-up. The terminal voltage U of the cell is the sum of following contributions:

$$U = U_0 + \sum |\eta| + R_0 I. \tag{2.1}$$

First contribution is U_0, the electromotive force of the electrochemical cell (given by the Nernst equation). The second contribution is the sum of the various over potentials and $R_0 I$ is the ohmic drop across the cell. At high current densities, the sum of the over potentials will, in a first approximation, be almost constant and sum up, together with U_0, to the water decomposition potential U_d. Thus, equation 2.1 can be rewritten as:

$$U = U_d + R_0 I. \tag{2.2}$$

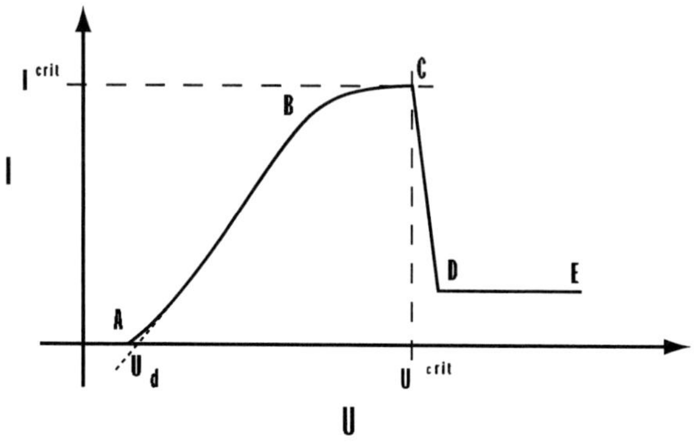

Figure 2.1. Typical current-voltage (I-U) characteristics of electrolytic water decomposition.

Note that in equation 2.2, R_0 is the inter-electrode resistance, function of the geometry and the electrolyte conductivity. If bubbles are evolving at the electrodes, the ideal law 2.2 will no longer be followed. The observed deviations are a good indication of the bubble activity. In order to study this effect on one electrode only (for example the cathode), it is important to choose a configuration in which the counter-electrode has a significantly larger surface. In this case, the current density at the counter-electrode will be much smaller than at the working electrode and so the bubble growth activity.

Figure 2.1 shows typical current-voltage characteristics of an electrochemical cell. Only up to point B the characteristics follows equation 2.2. Five different regions are distinguished (all numerical values are given for a working electrode of about 1 mm^2 and a counter electrode of a few centimeters square in a 30 $wt.\%$ NaOH solution):

Thermodynamic and over potential region : For a terminal voltage smaller than the water decomposition potential U_d ($U_d \approx 2$ V), no significant electrolysis happens and almost no current flows between the electrodes.

Ohmic region AB (typically from 2 to $10-12\ V$**)** : The mean current I increases nearly linearly with the terminal voltage U. Bubbles evolve at the working electrode and a compact bubble layer (with a thickness of a few millimeters) is formed. All bubbles have almost the same diameter of about 100 μm.

Limiting current region BC : The mean current I reaches a limiting value and is almost constant. In this voltage range, bubbles coalesce. Not only small bubbles, as in the ohmic region, but larger bubbles are also observed (typically from 300 to 700 μm). The mean temperature of the bulk electrolyte increases. The extent, and in general the presence, of this region depends on the wettability of the electrode [16, 17]. The end of the limiting current region is reached at the critical point C characterized by the critical voltage U^{crit} (typically around $20-30\ V$) and a critical current I^{crit} (typically around 1 A). Beyond this point, the current decreases very rapidly with the terminal voltage U.

Instability region CD : The mean stationary current decreases quickly with the terminal voltage U. The electrolyte resistance diverges and a gas film is formed around the working electrode. This region is called instability region because the system can be in a state similar to either the limiting current region BC or the electrochemical discharge region DE.

Electrochemical discharges region DE : The last region is characterized by the existence of a gas film around the electrode. The working electrode is completely isolated from the electrolyte. The mean current is low (in the milliampere range). The instant current can be described as a succession of short pulses (a few hundred microseconds with amplitude up to 1 A). Electrical discharges take place through the gas film.

2.2. Bubble Evolution at Gas Evolving Electrodes

2.2.1. Bubble Formation

During electrolytic decomposition of water, gas bubbles are produced at the electrodes. The formation of a bubble involves two steps. As a concrete example we discuss the formation of hydrogen bubbles. The first step is the microscopic formation of H_2 by an electrochemical process:

$$2H^+ + 2e^- \leftrightharpoons H_2. \tag{2.3}$$

In general this microscopic process is complex and involves several elementary steps, not all of them being necessarily electrochemical processes.

The second step is the macroscopic formation of bubbles by accumulation of the dissolved H_2 in the vicinity of the electrode:

$$H_2 \ (dissolved) \leftrightharpoons H_2 \ (bubble). \tag{2.4}$$

Whether or not gas bubbles will be formed depends on the local concentration of the dissolved hydrogen. At low current densities, the dissolved gas is removed from the electrode surface quickly enough, by diffusion and liquid convection, to avoid reaching sufficient supersaturation for bubble formation. For higher current densities, bubble growth starts at defects (such as cavities, pits and scratches able to contain some entrapped gas) on the electrode surface as now the electrolyte in the vicinity of the electrode is supersaturated with dissolved gas. During their growth, the bubbles are fed from the highly supersaturated surrounding electrolyte [18–20]. Two forces act on the bubbles: the buoyancy force F_B and the capillary force F_C:

$$F_B = \frac{4}{3}\pi R_b^3 \rho g \tag{2.5}$$

$$F_C = 2\pi D \gamma \sin(\alpha) \tag{2.6}$$

where ρ is the density of the electrolyte, γ its surface tension, R_b is the bubble radius, D is the order of magnitude of the aperture of an active nucleation site, and α is the contact angle. Detachment of the bubble occurs if the condition $F_B = F_C$ is satisfied. After detachment, some gas remains on the nucleation site promoting the formation of another bubble at the same place.

From the detachment condition $F_B = F_C$ one may anticipate that the mean bubble departure radius is well defined for a given electrode-electrolyte configuration (typical values are around 50 μm [15, 21]). It may be expected that the mean bubble departure radius is mainly a property of the electrode (the electrode surface roughness influences D), the electrode wettability (the contact angle α), and the electrolyte (density and surface tension of the electrolyte). Actually, several external parameters influence the bubble diameter distribution such as concentration, pH of the electrolyte, polarity and potential of the electrode, wetting conditions, and the current density [15, 21].

2.2.2. Bubble Layers

The bubbles around a gas evolving electrode organize themselves in a layer structured in three regions [18, 22]:

- adherence region,
- bubble diffusion region, and
- bulk region.

In the adherence region the bubbles adhere to the electrode surface and grow. The typical thickness of this layer is the mean bubble diameter. During the growing phase coalescence of bubbles occurs if they have mechanical contact [15]. After leaving the electrode surface the bubbles diffuse into the bubble diffusion region. This region is characterized by a high bubble concentration [18] and a typical thickness of a few millimetres. The bubbles can still grow.

The third region is the bulk region, which contains only a few dispersed bubbles. It is generally acknowledged that in this region the gas supersaturation is close to the level of the bulk solution and therefore bubbles do no longer grow. Sometimes large bubbles jumping directly from the adherence region to the bulk region are observed [22].

Each of these three regions affects the current-voltage characteristics of the electrochemical cell. During their growth in the adherence region, the bubbles partially shadow the active surface of the electrode. As the covered fraction is electrochemically inactive, the local current density j_{local} is increased compared to the nominal current density $j_n = I/A$, where I is the current flowing through an electrode of surface A:

$$j_{local} = \frac{1}{1-\theta}\frac{I}{A}, \tag{2.7}$$

with θ, the bubble coverage fraction defined as the fraction of the electrode area shadowed by the bubbles [23–25].

The presence of the bubble diffusion region generates an additional potential drop in the electrochemical cell compared to equation 2.1:

$$U = U_0 + \Sigma|\eta| + R^{bulk}I + R^{diff}I, \tag{2.8}$$

where R^{bulk} is the resistance of the bubble bulk region and R^{diff} the resistance of the diffusion region. The diffusion region resistance can be estimated by computing the electrolyte conductivity κ_{eff} of the electrolyte, using for example Bruggeman equation:

$$\kappa_{eff} = \kappa(1-\varepsilon)^{3/2}, \tag{2.9}$$

with ε the gas void fraction in the diffusion region and κ the electrical conductivity of the electrolyte without bubbles. For example, if the bubble diffusion region is modeled as a cylinder of thickness d and height h around the electrode of radius b, one can write:

$$R^{diff} = \frac{1}{(1-\varepsilon)^{3/2}} \frac{1}{2\pi\kappa h} \ln\left(\frac{b+d}{b}\right). \tag{2.10}$$

Note that in this description the shadowing of the surface by the bubbles in the adherence region was not considered. With increasing bubble coverage fraction θ, the resistance in the diffusion region will also increase. A possible ansatz is to write equation 2.11 as:

$$U = U_0 + \sum |\eta| + R^{bulk} I + \frac{R^{diff}}{1-\theta} I. \tag{2.11}$$

The estimation of the bubble coverage fraction θ is discussed in the following section.

2.2.3. Bubble Coverage Fraction

Depending on their size, the bubbles may be able to leave the adherence region or not. In order to describe quantitatively the bubble size distribution, an idealized model of the nucleation sites on the electrode is introduced [26]. The geometry of the model is depicted in figure 2.2. The electrode of radius b and height h is dipped into the electrolytic solution. The lateral electrode surface is subdivided into a lattice. Each lattice site represents a nucleation site where a bubble may grow. The probability that a bubble grows at a given site is equal to θ, the bubble coverage fraction. A site with a growing bubble is called an active site. Clusters (a group of neighboring active sites) are interpreted in this model as coalesced bubbles.

In order to derive an equation describing the evolution of the bubble coverage fraction θ for gas evolving electrodes, we will write a mass balance equation between the amount of gas produced and the quantity of gas leaving the adherence region [3, 27, 28]. Using Faraday's law one computes $\dot{V}_G$, the amount of gas produced per unit time at an electrode:

$$\dot{V}_G = \beta I, \tag{2.12}$$

with β the coefficient of faradic gas generation:

$$\beta = \frac{RT}{z\nu F p}. \tag{2.13}$$

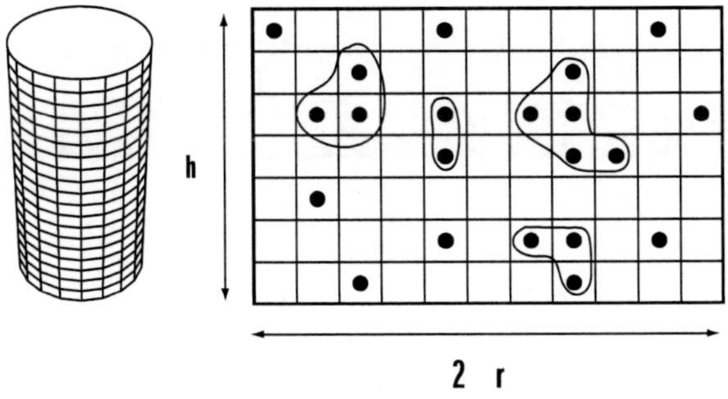

Figure 2.2. Percolation model of the bubble adherence region: the lateral electrode surface is subdivided into a lattice. Sites are occupied with a mean occupation probability θ. Neighboring bubbles are assumed to coalesce and form a new larger bubble [26].

Here R is the universal gas constant, T is the temperature at the electrode surface, F is Faraday's constant, p is the pressure, z is the charge number, and ν is the stoichiometric number. Together with equation 2.7, one obtains:

$$\dot{V}_G = \beta j_{local}(1-\theta)A. \tag{2.14}$$

The amount of gas leaving the electrode surface is given by the mean number of bubbles leaving the electrode per unit time. To evaluate this number, we have to count all bubbles that are able to quite the adherence region. For each bubble coverage fraction θ is associated a distribution of cluster of size s. Percolation theory allows to describe this distribution by specifying the number of s-clusters per lattice site $n_s(\theta)$.

We assume that only bubbles of size smaller than s^{max} can leave the electrode surface. The value of s^{max} is a function of the electrode geometry and wettability. The fraction of nucleation sites occupied by growing gas bubbles able the leave the electrode is given by $\sum_{s=0}^{s^{max}} s n_s(\theta)$. If V_b is the mean volume of a bubble growing on a nucleation site at the moment of coalescence with its neighbors, N is the mean number of nucleation sites per electrode surface, and Δt_b^{-1} is the mean bubble detachment frequency, then the mean total volume of

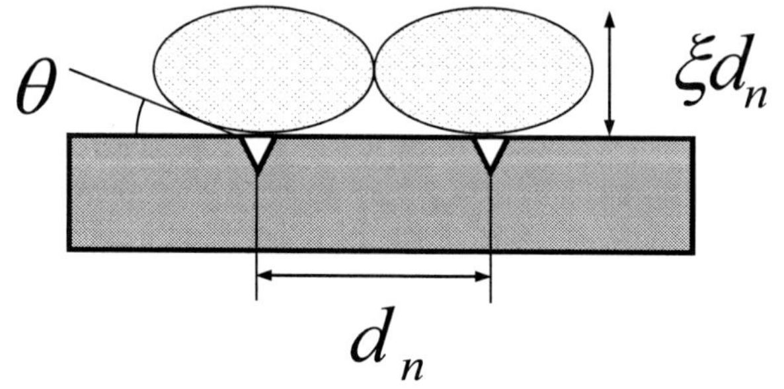

Figure 2.3. Geometry of two coalescing bubbles in the adherence region.

gas leaving the electrode per unit of time and surface is:

$$\dot{V}_D = \frac{V_b}{\Delta t_b} NA \sum_{s=0}^{s^{max}} s n_s(\theta). \tag{2.15}$$

Using equations 2.14 and 2.15 one can directly write the following mass balance equation:

$$\frac{d}{dt}(V_b NA\theta) = \dot{V}_G - \dot{V}_D = \beta j_{local}(1-\theta)A - \frac{V_b}{\Delta t_b} NA \sum_{s=0}^{s^{max}} s n_s(\theta). \tag{2.16}$$

Note that the product $V_b NA$ is the volume occupied by the gas bubbles if all nucleation sites would be active at the same time. In order to take into account the wettability of the electrode surface (figure 2.3), we write the volume of a single isolated bubble as $V_b = \sigma_b \xi d_n$, where $\sigma_b = 1/N$ is the surface occupied by this bubble and ξd_n is its height written as the product of d_n (the mean distance between two nucleation sites) and ξ (the parameter describing the degree of flatness of the bubble). The parameter ξ is function of the wettability and the electrode geometry. In particular, for small cylindrical electrodes, where the electrode radius is similar to the bubble radius, ξ can typically reach values around 2. Using these definitions, equation 2.16 is written as:

$$\frac{d\theta}{dt} = \frac{\beta j_{local}}{\xi d_n}(1-\theta) - \frac{1}{\Delta t_b} \sum_{s=0}^{s^{max}} s n_s(\theta). \tag{2.17}$$

The solution of the differential equation 2.17, together with equation 2.11, allows the computation of the current-voltage characteristics of an electrochemical cell with gas evolving electrodes.

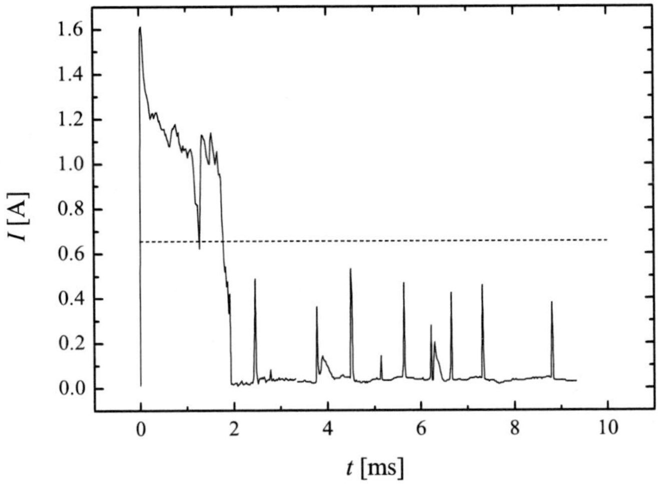

Figure 2.4. A typical example of a voltage step input of 30 V. The working cathode has a diameter of 0.17 mm and the electrolyte used is 30 $wt.\%$ NaOH solution [26].

2.3. Gas Film Formation

As described in section 2.1., a gas film forms around the working electrode passed beyond the critical point characterized by U^{crit} and I^{crit}. Experimentally the formation of the gas film can be followed by observing the current flowing in the system. Figure 2.4 shows a typical example. A 0.17 mm stainless steel cylinder cathode is dipped in a 30 $wt.\%$ NaOH solution. A voltage step of 30 V is applied. One can clearly see how after 2 ms the gas film is formed and the first discharges appear. Two main mechanisms are responsible for the formation of the gas film: local evaporation of the electrolyte due to joule heating and coalescence of the electrochemically formed bubbles on the electrode surface. A combination of both processes is also possible.

2.3.1. Local Electrolyte Evaporation

The formation of gas film around a thin electrode happens if locally the temperature reaches the electrolyte evaporation temperature. If the heat dissipation by thermal conduction through the electrode and the electrolyte is neglected, the heat Q brought to the electrolyte of mass m and heat capacity c will increase its temperature by ΔT according to:

$$Q = mc\Delta T. \tag{2.18}$$

The heat provided to the electrolyte in the time interval Δt comes from joule heating by the current I:

$$Q = RI^2\Delta t, \tag{2.19}$$

where R is the resistance of the electrolyte. A gas film can be built if a thin shell around the electrode reaches the evaporation temperature T_e of the electrolyte. Assuming a radial current distribution around the electrode of radius b and height h, one derives easily from equations 2.18 and 2.19, the following expression for the time t_f needed to build a gas film by local electrolyte evaporation [3, 29]:

$$t_f = \frac{\kappa\rho c}{j_n^2}(T_e - T_0), \tag{2.20}$$

where $j_n = I/(2\pi bh)$ is the nominal current density at the electrode surface and T_0 is the bulk temperature of the electrolyte. Figure 2.5 shows, for typical values of an electrolyte ($\kappa = 0.4\ \Omega^{-1} \cdot cm^{-1}$, $c = 4.18 \cdot 10^3\ J \cdot kg^{-1} \cdot K^{-1}$, $\rho = 10^3\ kg \cdot m^{-3}$), the estimation of the gas film formation time as a function of the nominal current density. Equation 2.20 underestimates the gas film formation time as the heat transport through the electrolyte is neglected. A more precise estimation can be obtained by taking into account the heat dissipation inside the electrolyte. Therefore, Fourier equation has to be solved. Numerical calculations [3] show that equation 2.20 is an excellent approximation to the problem.

Note that the gas film formation time is still underestimated as at least three important effects are neglected: the heat dissipation through the electrode, heat loss by convection and the heat needed for the evaporation of the electrolyte (latent heat).

In 1902, Klupathy [29] derived equation 2.20. By comparing his calculations with experimental data (for a nominal current density of about $0.6\ A \cdot mm^{-2}$, he measured gas film formation times of 4 ms), he concluded that the joule heating inside the electrolyte cannot account for the gas film formation for the considered geometries. More recently, Guilpin [30] and Fascio [16] found, for different geometries than those considered by Klupathy, similar gas film formation times as the one predicted by equation 2.20.

The conclusion to be drawn here is that this simplified model indicates that in principle it is possible to build the gas film by local electrolyte evaporation

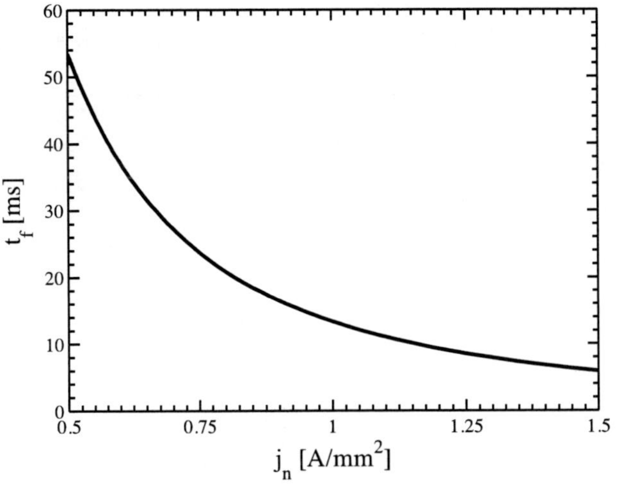

Figure 2.5. Estimation of the gas film formation time t_f if only local joule heating is considered (heat dissipation is neglected); computed using equation 2.20 with $\kappa = 0.4\ \Omega^{-1} \cdot cm^{-1}$, $c = 4.18 \cdot 10^3\ J \cdot kg^{-1} \cdot K^{-1}$, $\rho = 10^3\ kg \cdot m^{-3}$. Reprinted from [3] with permission from Elsevier.

by joule heating, even it seems likely that other processes must be involved as well.

2.3.2. Electrochemical Gas Evolution

At high current densities, bubbles coalesce to form larger bubbles. It becomes possible to create a large enough structure on the electrode surface such that the buoyancy force can no longer overcome the adherence force of the coalesced bubble. A gas film forms at the electrode surface. Recall that the evolution of the bubbles growing on the electrode surface is given by equation 2.17, which writes for the stationary situation:

$$\frac{\beta j_{local}}{\xi d_n}(1-\theta) = \frac{1}{\Delta t_b} \sum_{s=0}^{s^{max}} s n_s(\theta). \tag{2.21}$$

As soon as a bubble of size larger than s^{max} appears, the formation of the gas film starts. If we consider the approximation $s^{max} = \infty$ (a bubble of similar size as the electrode), the formation of the gas film starts at the critical current I^{crit}

Table 2.1. Percolation thresholds p_c for some lattices [31].

Lattice	p_c
Honeycomb	0.696
Square	0.593
Triangular	0.5

[3]:

$$\frac{I^{crit}}{A} = \frac{\xi d_n}{\beta \Delta t_b} p_c, \tag{2.22}$$

where p_c is the bubble coverage fraction at which a cluster of infinite size appears. The values of p_c are function of the lattice geometry of the nucleation sites considered and can be determined in the framework of percolation theory (see table 2.1 for some values [31]). Equivalently, the gas film starts to grow at the critical voltage U^{crit} given by [3]:

$$U^{crit} = U_d + R(p_c) A \frac{\xi d_n}{\beta \Delta t_b} p_c, \tag{2.23}$$

where $R(p_c)$ is the inter-electrode resistance at the moment of the gas film creation. Some typical order of magnitudes can be computed. If one considers a cathode with hydrogen evolution, then $\beta = 0.13\ cm^3 \cdot A^{-1} \cdot s^{-1}$. For a cylindrical cathode of a few hundred micrometers in diameter, it is known experimentally that $\Delta t_b \approx 1\ ms$ and $\xi d_n \approx 0.2\ mm$ [28]. It follows that $j^{crit} \approx 1\ A \cdot mm^{-2}$ (with $p_c \approx 0.5$) and that $U^{crit} = 20 - 30\ V$, depending on the value of $R(p_c)$.

Figure 2.6 shows the effect of the electrolyte concentration on the critical voltage and current. The electrolyte concentration affects the electrolyte electrical conductivity, but as well the parameters ξd_n (wettability) and Δt_b (as the concentration affects the viscosity and density of the electrolyte). The decrease in U^{crit} with the electrolyte concentration is not specific to NaOH but a generally observed effect.

Equations 2.22 and 2.23 capture several important parameters responsible for the formation of the gas film. The following scenarios for its formation result [33]:

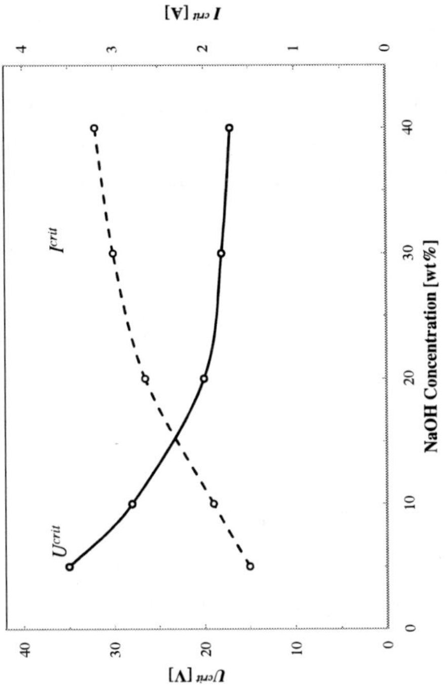

Figure 2.6. Critical voltage U^{crit} and critical current I^{crit} as a function of the electrolyte concentration for a 1.1 *mm* cylindrical electrode 2 *mm* immersed in a aqueous solution of NaOH (Plotted according to data from [32]).

- Increase in the nominal current density: With increasing nominal current density I/A, the bubble coverage fraction increases. When it reaches a critical value, as computed by equation 2.22, a gas film is formed. Such a scenario happens, in current-controlled cells or in voltage-controlled cells if for some reasons the inter-electrode resistance decreases.

- Deteriorated wetting: The wetting of the electrode directly influences the bubble shapes (through the parameter ξ). For a given bubble volume, the shadowed surface on the electrode will increase with deteriorated wetting [34]. Several parameters affect the wetting such as the electrode surface roughness and porosity and also the local over potential. Changing the wetting of the electrode-electrolyte-gas interface can be used to change significantly the critical voltage [28].

- Decreasing rate of bubble release: The mean bubble detachment frequency $1/\Delta t_b$ directly controls the onset of the gas film. For large enough bubbles, the bubble detachment time Δt_b becomes so large that the gas film can be formed. Δt_b is affected by other parameters such as the wetting of the electrode, viscosity and density of the electrolyte, or the local hydrodynamic fluxes.

- Increased rate of gas production: The rate of bubble formation is directly controlled by the local current density. Note that the local current density can be only indirectly controlled in experiments using the nominal current density or the terminal voltage. The formation of vapor by local joule heating can also contribute to this increase.

The time needed to form a gas film by bubble coalescence can be estimated by solving equation 2.17 for a voltage step input. By defining the normalized voltage $\mathcal{U}$ as:

$$\mathcal{U} = \frac{U - U_d}{U^{crit} - U_d}, \tag{2.24}$$

one finds the following expression for the gas film formation time t_f [3]:

$$\frac{t_f}{\Delta t_b} = \frac{1}{1 + \frac{p_c}{1 - p_c}} \ln\left[\frac{\mathcal{U}}{(1 - p_c)(\mathcal{U} - 1)}\right]. \tag{2.25}$$

2.4. Gas Film Dynamics

The gas film formed around an electrode under very high current density can be described, according to Kellogg, as a vibrating structure [35]. Kellogg also stressed out that once the gas film is formed, no further bubble generation can be observed. He concluded that the electrochemical reactions must now take place at the gas-electrolyte interface. Another important finding by Kellogg is that the thickness and uniformity of the gas film is a function of the electrolyte's bulk temperature. At higher temperatures the gas film becomes more uniform, vibrates only slightly, and is thinner. Han et al. [36] showed that the gas film dynamics can as well be influenced by adding ultrasonic vibrations to the electrolyte.

In connection with Kellogg's contribution, recent measurements (Figure 2.7) done by the authors [37] show that the gas film life time t_g increases with

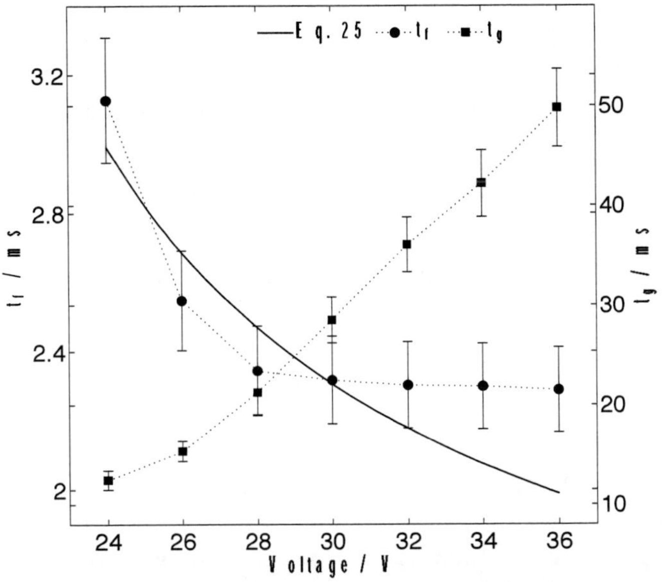

Figure 2.7. Gas film life time t_g and gas film formation time t_f versus the cell terminal voltage. The solid line is plotted according to equation 2.25 with $U^{crit} = 21\ V$, $U_d = 2.2\ V$ and $\Delta t_b = 2.33\ ms$. The dashed line is plotted according to equation 2.20 with $\kappa = 29.2\ S \cdot m^{-1}$, $\rho = 1330\ kg \cdot m^{-3}$, $U_d = 2.2V$, $T_e = 100°\ C$, $T_0 = 20°\ C$ and $R = 2.3\ \Omega$. Reprinted from [37] with permission from Elsevier.

the cell terminal voltage whereas its formation time t_f deceases as predicted by equation 2.25. The increase of the gas film stability with the cell terminal voltage is probably linked to the increase of the electrode temperature. As measured by Kellogg [35], the electrode temperature increases more or less proportionally with the cell terminal voltage.

In a recent paper by El-Haddad and Wüthrich [38], a stochastic model which complements the temporal evolution of the bubble coverage fraction (Equation 2.17) was proposed. The model is based on the assumption that the gas film detaches from the electrode surface following a poisson process. With their work, the authors proposed another way to study the gas film dynamics rather than the direct method of current timeseries processing [37]. This is by reformulating equation 2.17 as follows [38]:

$$\frac{d\theta}{dt} = \frac{\beta j_{local}}{\xi d_n}(1-\theta) - \frac{1}{\Delta t_b}\sum_{s=0}^{s^{max}} s n_s(\theta) - \frac{1}{\Delta t_g} P(\theta), \quad (2.26)$$

where $P(\theta)$ is the size of the infinite cluster and $1/\Delta t_g$ represents the frequency at which the gas film detaches from the electrode surface. Equivalently, one can write the dimensionless version of equation 2.26 as follows:

$$\frac{d\theta}{d\bar{t}} = \frac{p_c}{1-p_c}\mathcal{U}(1-\theta) - \sum_{s=0}^{s^{max}} s n_s(\theta) - \frac{1}{\Delta \bar{t}_g} P(\theta). \quad (2.27)$$

In this general equation, $\bar{t} = t/\Delta t_b$ and $\Delta\bar{t}_g = \Delta t_g/\Delta t_b$ represent the normalized time and the normalized average gas film detachment time by the average bubble detachment time, respectively. The stationary solutions of equation 2.27 for various gas film detachment times $\Delta\bar{t}_g$ are multiple (stable and unstable solutions) and show an increase of the unstable region of the current-voltage characteristics with increase of $\Delta\bar{t}_g$. From the solutions of the time-dependent equation at a fixed value of $\Delta\bar{t}_g$, the authors showed an enlargement of the bistable region with the increase of the scan rate. This is an important result and is of particular practical use to estimate the dynamics of the gas film in the electrochemical discharge phenomenon (see figure 2.8). This can be simply done by performing onward and backward scans and calculate the width of the hysterisis by the following formula:

$$\Delta\mathcal{U} \approx \bar{v}\cdot\Delta\bar{t}_g, \quad (2.28)$$

with $\Delta\mathcal{U}$ the extension of the hysteresis. As a further result it is shown that the extension of the limiting current region $\Delta\mathcal{U}^{BC}$ is related to the variance $\mathrm{var}(\Delta\bar{t}_g)$ of the stochastic gas film detachment process as:

$$\Delta\mathcal{U}^{BC} \approx \bar{v}\cdot\mathrm{var}(\Delta\bar{t}_g) \quad (2.29)$$

2.5. Discharge Activity

The formation of a gas film around an electrode drastically modifies the potential distribution. When a gas film is present, the mean current is very small and there is nearly no ohmic drop in the electrolyte. As a consequence, almost all

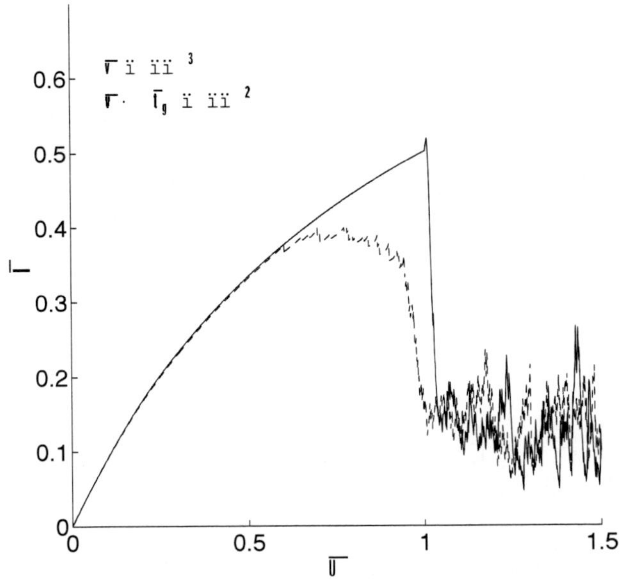

Figure 2.8. Dimensionless time-dependent solution of the normalized bubble evolution equation. With kind permission from Springer Science+Business Media [38].

the potential drop takes place across the gas film resulting in quite high local electrical fields responsible for electrochemical discharges. The exact nature of these discharges is still open to discussion. In the case of a gas film formed around an anode, the so-called Hickling and Ingram mechanism [39] is responsible of the discharge activity. This situation will be discussed in more details in chapter 3.

In case of a gas film formed around a cathode, Wüthrich [3, 26] proposed that electrochemical discharges are arc discharges. The initiation mechanism is probably thermal-assisted field emission of electrons (the electrode temperature is around $500°C$ [35]). The discharges can be well modeled by a Poisson process. Interestingly, two qualitative different behaviors in the discharge activity in function of the terminal voltage can be observed. For voltages slightly above the critical voltage (for voltages in the instability region), pulses of a few $100\ \mu s$ duration together with short pulses of around $10\ \mu s$ are observed (Figure 2.9 top). The number of discharges decreases with increasing terminal voltage. Above $32\ V$ only short pluses of about $10\ \mu s$ are observed (Figure 2.9 bottom).

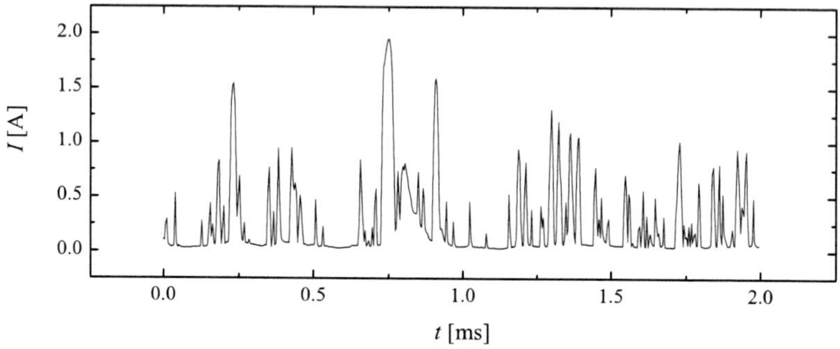

(a) 25 V

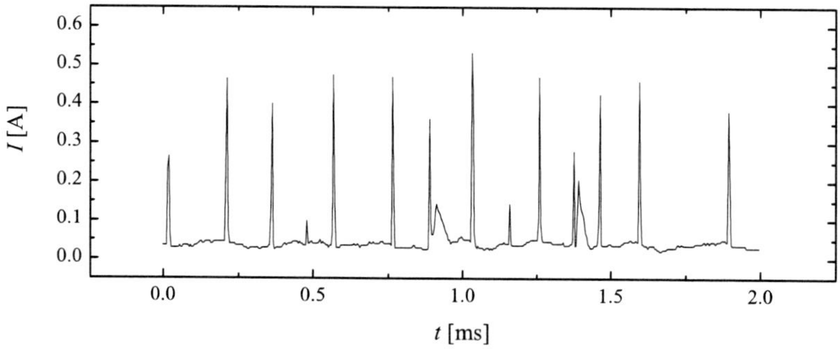

(b) 30 V

Figure 2.9. Typical current pulses for terminal voltages of 25 V (a) and 32 V (b) in the case of a gas film formed around a thin cylindrical cathode in NaOH solution [26].

The number of discharges increases according a field emission law with the terminal voltage [3, 26]. The voltage at which the transition to the field emission behavior takes place is sometime called the transition voltage [40].

Chapter 3

Electrochemical Discharges

The electrochemical discharge phenomenon occurring inside a gas film formed around a working electrode presents features very similar to radiative chemistry of water [41]. Back to 1964, Hickling and Ingram [39] have already stated a qualitative comparison between glow discharge electrolysis and α-radiolysis of aqueous solutions. We will therefore summarize some general experimental and theoretical results of linear energy transfer in radiation chemistry of water followed by the chemical and reactional mechanisms at the microscopic level. General statements about electrode sputtering phenomenon are also included. This chapter is crucial for the understanding of the several applications associated with the different electrochemical processes covered by this book.

3.1. Reaction Mechanisms

3.1.1. Radiation Chemistry

The current state of knowledge of radiation chemistry of aqueous solutions containing the ions M is summarized in its general reactional mechanism as illustrated in figure 3.1 [42]. It also includes the time scale at which the various reactions are happening. The first reaction with this process gives birth to the ionized state of the water molecule, H_2O^+ and one electron e^-, in addition to the electronically excited state, H_2O^*:

$$H_2O \rightarrow H_2O^+ + e^- \text{ and } H_2O^*. \tag{3.1}$$

The so-called ion-molecule reaction of the excited ionic state H_2O^+ with water

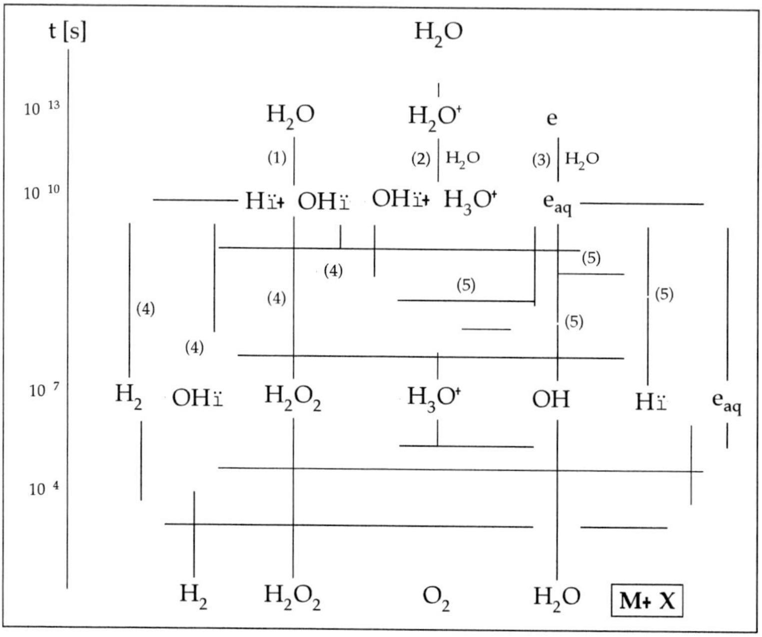

(1) Dissociation; (2) Ion-molecule; (3) Solvation; (4) Dimerization; (5) Recombination.

Figure 3.1. Water irradiation mechanism [42].

molecule gives both OH radical and the hydronium cation H_3O^+ such as:

$$H_2O^+ + H_2O \rightarrow OH\ + H_3O^+, \tag{3.2}$$

while the dissociation of the H_2O^* gives the H and OH radicals, with the possible reaction given H_2 and O , as follows:

$$H_2O^* \rightarrow H\ + OH\ ,\ H_2 + O\ . \tag{3.3}$$

On the other hand, within this series of ultra fast reactions occurring in the times span of the electronic transition, e^- will undergo a solvation reaction. With water as the solvent, the solvation or more specifically the hydration reaction is written as:

$$e^- + nH_2O \rightarrow e^-_{aq}. \tag{3.4}$$

The solvated electron was for the first time observed by transient optical absorption measurements by both Hart and Boag [43] in the year 1962. It is

Table 3.1. Dimerization and recombination reactions, adapted from Buxton [41].

Chemical reaction	Reaction rate constant k †
$e^-_{aq} + e^-_{aq} \longrightarrow H_2 + 2\,OH^-$	0.54
$e^-_{aq} + OH \longrightarrow OH^-$	3.0
$e^-_{aq} + H_3O^+ \longrightarrow H + H_2O$	2.3
$e^-_{aq} + H \longrightarrow H_2 + OH^-$	2.5
$H + H \longrightarrow H_2$	1.3
$OH + OH \longrightarrow H_2O_2$	0.53
$OH + H \longrightarrow H_2O$	3.2
$H_3O^+ + OH^- \longrightarrow 2\,H_2O$	14.3

† $k \cdot \backsim 10^{10}\,M^{-1} \cdot s^{-1}$.

a transient chemical species where the electron e^- is surrounded by solvent molecules. This trapping process happens once the electron reaches its thermal equilibrium with the liquid, which is achieved in less than the picosecond time scale. Hart and Boag have successfully produced the solvated electrons, or more specifically here the hydrated electrons, by means of pulse radiolysis of aqueous solutions. Other ways to produce the e^-_s are known [41, 44], such as chemical and electrochemical methods as well as photolysis using flash lamps or lasers. As far as it concerns the chemical reactivity of the solvated electron, and in spite of its short life-time, it is considered as a strong reducing element. Recombination of the e^-_s with positive ions and radicals are produced inside volume elements called spurs, in radiation chemistry.

In a second step, depending on their kinetic energies, the products H , OH , H_3O^+ and e^-_{aq} can react together by dimerization and recombination reactions as well as off themselves. Table 3.1 lists the possible reactions at this step where the ones involving the cations M are shown in figure 3.1 with dashed lines. From results of water radiolysis at $0.23\ eV \cdot nm^{-1}$, products from radiated water molecule, up to the 10^{-7} s time period, can be described by the following overall reaction [41]:

$$H_2O \rightsquigarrow 0.28\ e^-_{aq},\ 0.063\ H\ ,\ 0.28\ OH\ ,\ 0.047\ H_2, \\ 0.073\ H_2O_2,\ 0.28\ H_3O^+, \quad (3.5)$$

where the numbers represent the radiation chemical yield in 10^{-6} mol/J, reported generally in terms of the G-values. $G(X)$ is known as the number of molecules of product X formed per 100 eV energy absorbed. Note that commonly the unit molecule/100 eV is used where 1 mol J^{-1} equals to $9.6490 \cdot 10^6$ molecule/100 eV. One can remark that the main chemical yields in reactions 3.5 are the hydrated e^-_{aq}, the hydroxyl radical OH and the hydronium cation H_3O^+, where the reactions give about 2.7 molecule/100 eV of each of these products. However, at much higher radiative energies, i.e. 108 $eV \cdot nm^{-1}$, chemical yields in reactions 3.5 becomes as follows [41]:

$$H_2O \rightsquigarrow 0.044\ e^-_{aq},\ 0.028\ H\ ,\ 0.056\ OH\ ,\ 0.11\ H_2, \\ 0.11\ H_2O_2,\ 0.044\ H_3O^+, \quad (3.6)$$

where now the high yield is for the molecular hydrogen and hydrogen peroxide with $G(H_2) = G(H_2O_2) \approx 1.1$ molecule/100 eV, due to overlapping spurs [41].

These transient reactions are then followed in time by several other reactions due to the diffusion of the products toward the bulk. Several molecular and secondary radical products are obtained such as H_2, H_2O_2, O_2, H_2O and stable products of reactions with metallic ions M present in the aqueous solution [41]. These later reactions given reduced or oxidized states of M are due to reactions with H_2, OH , H_2O_2, H and e^-_{aq}. The solvated electron e^-_{aq} is the strongest reducing agent obtained in the aqueous solution, with a standard reduction potential with respect to the standard hydrogen electrode (SHE) of $E°(H_2O/e^-_{aq}) = -2.87\ V$ as well as the H atom which is a powerful reductant agent in acidic solution with $E°(H^+/H) = -2.31\ V$, following reaction 3.7 [41] as:

$$e^-_{aq} + H^+ \longrightarrow H\ . \quad (3.7)$$

3.1.2. Glow Discharge Electrolysis

Glow discharge electrolysis (GDE) occurs when an electric discharge passes from an electrolyte to an electrode situated above the solution. This can be realized for thin electrodes placed in the vicinity of an electrolyte (with a gap of a few millimeters). The voltages needed are typically a few hundred volts. Since the pioneering work of Gubkin [45], it is known that glow discharge electrolysis can be used to reduce metallic salts in an electrolytic solution. One of the main products of anodic glow discharge electrolysis is hydrogen peroxide, but surprisingly the yield is greater than predicted by Faraday's law. This is explained by Hickling and Ingram [39] by two processes: a polar and an apolar process. The polar process corresponds to the charge transfer reaction (electrochemical reaction). The apolar process is responsible for oxidation accompanied by the evolution of hydrogen from the solution (chemical reaction). According to this model, H_2O^+ ions are accelerated by the high electrical field in the gap between the anode and the electrolyte through the gas phase. These ions can have energies up to several hundred electron volts depending on the voltage used. The presence of H_2O^+ inside the gas phase in the vicinity of the anode is possible considering the high temperature below the glow spot that is able to vaporize the water from the electrolyte. The accelerated ions penetrate the liquid with energies that are sufficiently high to ionize or activate water molecules from the electrolyte by collision:

$$H_2O^+ + H_2O \rightarrow H_2O^+ + e^- + H_2O^+ \quad : \quad \text{Ionization} \tag{3.8}$$

$$H_2O^+ + H_2O \rightarrow H_2O^* + H_2O^+ \quad : \quad \text{Activation} \tag{3.9}$$

$$H_2O^* \rightarrow H + OH \tag{3.10}$$

These reactions take place in a small region (a thickness of about 10 Å, estimated based on analogy with radiation chemistry) of the solution, which is termed the primary reaction zone. The H_2O^+ may also undergo the charge transfer reaction:

$$H_2O^+ + H_2O \rightarrow OH + H_3O^+ . \tag{3.11}$$

The final products of processes 3.8 to 3.11 are free OH radicals. These radicals can interact and produce hydrogen peroxide. In summary, the reactions taking place during anodic glow discharge electrolysis are:

$$2H_2O^+ \rightarrow H_2O_2 + 2H^+ \tag{3.12}$$

$$2H_2O^* \rightarrow H_2O_2 + H_2 \tag{3.13}$$

This mechanism, called the Hickling and Ingram mechanism, explains the formation of hydrogen peroxide and hydrogen during anodic glow discharge electrolysis. It also explains the observed higher yield than predicted by Faraday's

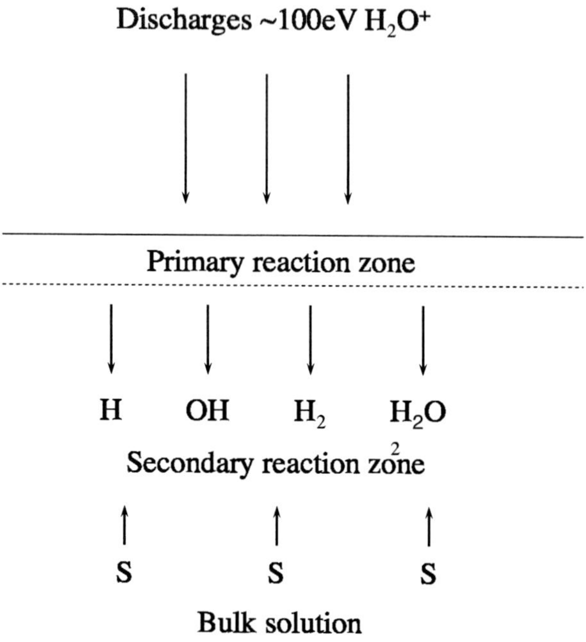

Figure 3.2. Mechanism of glow discharge electrolysis according to Hickling and Ingram [39].

law. Not all OH radicals will react. Consequently, the primary reaction zone is a mixture of H , OH , H_2O_2, and H_2. These products will diffuse towards the bulk electrolyte forming a secondary reaction zone (Figure 3.2). In this zone the diffusion products can undergo oxidation and reduction reactions with the components of the electrolyte.

3.1.3. Contact Glow Discharge Electrolysis

When electrochemical discharges occur thought a gas film formed around an electrode, one terms the phenomenon contact glow discharge electrolysis (CGDE). Depending on the polarization of the electrode, one distinguishes between anodic and cathodic contact glow discharge electrolyses. Both are discussed below. As for glow discharge electrolysis, in anodic contact glow discharge electrolysis hydrogen peroxide is synthesized at a rate higher than that predicted by Faraday's law [10]. An extended model of glow discharge electrolysis, proposed by Sengupta et al. [46], explains this observation. Two reaction zones are distinguished: the first inside the gas film and the second in the liquid phase near the gas film-electrolyte interface. The authors proposed that the electrochemical discharges inside the gas film will dissociate the water vapor into hydrogen and oxygen (the later being in excess to Faraday's law) by the following mechanism:

$$H_2O \rightarrow H + OH \tag{3.14}$$
$$H + H \rightarrow H_2 \tag{3.15}$$
$$OH + OH \rightarrow H_2O + \frac{1}{2}O_2 \tag{3.16}$$
$$H + OH \rightarrow H_2O \tag{3.17}$$

This mechanism is consistent with the observation of H lines and OH bands in the emission spectra of electrochemical discharges at an anode [30, 42].

In the liquid phase, Sengupta et al. [46] proposed a mechanism that is very similar to the Hickling and Ingram mechanism, forming H_2O_2 and excess oxygen by the bombardment of H_2O^+. In summary:

1. Traditional electrolytic charge transfer occurs at the gas film-electrolyte interface forming oxygen according Faraday's law (0.25 $mol/mol\ electron$).

2. Decomposition of water vapor inside the gas film by the electrochemical discharges by energy transfer resulting in the formation of hydrogen and excess oxygen.

3. The formation of H_2O_2 in the liquid phase by bombardment of H_2O^+. This mechanism produces non-Faradaic yields of H_2, H_2O_2, and O_2. Depending on the nature of the electrolyte, further chemical reactions may take place in the secondary reaction zone as in glow discharge electrolysis.

Cathodic contact glow discharge electrolysis presents some major differences. Physically, the gas film formed during the process is less stable when the working electrode is a cathode than in the case of an anode [35]. According to Hickling and Ingram [10] this effect may be attributed to the different thermal transfer processes. In the case of an anode, most of the heat produced is believed to be transferred to the electrolyte, helping in the formation of a stable gas film, whereas in the case of a cathode, most of the heat is removed through the electrode. Another significant difference is the nature of the discharges. In the case of a cathode, the electrochemical discharges are most likely electrons emitted from the electrode by thermionic emission, whereas in the case of an anode, the discharges result from the accelerated ions across the gas film.

The chemical effects in anodic and cathodic contact glow discharge electrolysis are similar [39]. The major difference is the reduced rate of the various reactions taking place. In particular, no H_2O_2 production is observed. Nevertheless, O_2 and OH radicals can be produced by the bombardment of the electrolyte-gas film interface by the electron streams. In contact glow discharge electrolysis charged species can acquire relatively high energies (compared to more standard electrochemical techniques). In a certain sense, CGDE is a simple technique to produce a low temperature plasma [42, 47–50]. The reactions that may take place in the secondary reaction zone can be used to produce a great number of chemical reactions such as synthesis of amino acids [11–13] or the degradation of organic components such as phenols [51–53], benzoic acids [49] and dyes [54]. Due to the formation of highly active species, such as OH radials, H radicals and H_2O_2, contact glow discharge electrolysis appears as an inserting candidate for waste water treatment. Another interning and promising application is in the field of surface treatment [2].

3.2. Physical Mechanisms

Through a series of experiments, Grove [55] demonstrated in 1852 the possible disintegration of negatively polarized electrodes by cation impacts using gas discharges. The sputtered materials from the cathode were then deposited on the inner wall of the discharge tube. This observation led to numerous experimental investigations and theoretical attempts to explain the phenomenon that was initially an undesired effect as the electrode material is destroyed. A historical overview, from the pioneering works of Kingdon and Langmuir in 1923 up to Thompson in 2004, among others, is presented.

In 1923, experimental evidence was presented by Kingdon and Langmuir [56] of a special case of sputtering, when they bombarded thoriated tungsten with different kinds of ions in a glow discharge tube. The authors remarked during their experiments the disintegration of the thin film of thorium on the tungsten substrate rather than the tungsten itself. In addition, they concluded that the sputtering rate, which represents the number of atoms removed by incident ion impact, increases with the incident ion mass and ion energy. Kingdon and Langmuir proposed the so-called momentum transfer model, an ejection mechanism for the electrode sputtering phenomenon stating that two successive bombardments on the same thorium atom of the thoriated tungsten filament are necessary for its disintegration. The first impact generates atomic size crater or depression at the filament surface while the second impact of the ion, in the bottom of this depression, removes one of the surrounding thorium atoms by reflection. The second impact succeeds in the thorium atom removal if the energy transferred is not too much dissipated so that it stays greater than the atomic heat of vaporization, which is the case according to the technical report of Langmuir [58], published a few years earlier, in 1916.

An excellent overview of the cathode sputtering phenomenon was given at that time by Massey and Burhop [59], from which some conclusions are retained. The first is that the apparent rate of sputtering at the electrode surface is the difference between the rate of sputtering due to the sputtered atoms and the rate of re-diffusion of these atoms back to the electrode. The effect of back diffusion, due to collisions in the gaseous phase, is function of the geometrical arrangement of the discharge tube as well as the pressure in the gaseous phase. The presence of oxide layer at the electrode and the electrode surface conditions are global parameters controlling the sputtering rate. In addition, the energy and the nature of the sputtering ions with which the electrode is bombarded, the nature of the sputtered material and the target temperature [60] play essential roles in the sputtering rates.

An important contribution in the field was done by Wehner [61], where he obtained pronounced high density areas or spots from sputtering single crystal electrodes. There was now strong evidence in favor of the momentum transfer theory proposed by Kingdon and Langmuir. The work of Henschke [62] reported in 1957 is based on mechanical collision theory taking into account oblique and normal incidence sputtering phenomena. In summary, the oblique impact sputtering involves the atoms belonging to the outermost plane of the electrode surface and therefore is a two-body collisional process. The incident ion penetrates the surface plane and hits the atom on its inside hemisphere to obtain an impulse momentum oriented outward normal to the surface.

The description of the more complex multi-body collisional process of ions normally incident to the electrode surface can be found in Henschke [62]. This is valid under the assumption that the atoms are binding to the surface of the electrode by a force normal to the surface. With an energy superior than the heat of vaporization, which corresponds to the binding energy necessary to the atom to adsorb on the surface, the atom is ejected. Minimum threshold energies for different contact angles and surface atom separations were calculated by Henschke assuming collisions. The ion starts a collision process with a lower surface atom where it gets reflected and hits, secondly, the lower hemisphere of an atom belonging to the outermost surface plane. Depending on the surface profile, triple-collision sputtering is possible according to Henschke [62].

Silsbee added to this collision model, that he called "the billiard ball model", the effect of the local geometry of the crystal [63]. The crystalline structure plays an important role in the successive collisions at the atomic layer of the electrode. From his two-body collision calculations, an atom is reflected from the surface in the direction of an angle β_i while for the subsequent collision, the angle β_{i+1} will decrease if the energy is low enough, and the collision cascade occurs along a close packed rows of atoms. This was called the "momentum focusing" concept which is nowadays the most widely accepted theoretical interpretation of the electrode sputtering mechanism.

Additional reviews and relevant works about sputtering and cluster formation are available elsewhere [57, 64–70], where in summary, it is to be retained that the sputtering yield is influenced by:

1. energy of incident ions,
2. incident angles of ions,
3. nature of target material,
4. crystal arrangement of target material.

The discharge regime during the electrochemical discharges is qualitatively similar to the material sputtering under ionic bombardment in gas discharges. Different researchers have pointed out this electrode erosion effect [71, 72]. According to these authors, the electrode is subject to local chemical etching process and, to a certain extent, to heating effects. Naturally, it is understood that the chemical etching rate is enhanced by the microscopic kinetic temperature generated at the electrode surface by the discharges themselves. The ion bombardment, during electrode sputtering phenomenon, unifies these explanations, by referring to the different parameters controlling the sputtering ratio from the overview of Massey and Burhop [59]. More specifically for the nanoclusters fabrication by electrochemical discharges, some of the ions present in the solution are accelerated toward the gas film due to the electric field and come to hit the atoms belonging to the outermost plane of the electrode. By this mechanism a two-body collision generates the ejection of an atom in gaseous phase. Many-body collisional concept is possible [62], where still the atom of the outermost plane which is ejected by collision of the ion already underwent several collisions before. The ions of this mechanism are mostly the metal cations present is the solution either from the dissolution dynamics of the anode or from the solute.

Chapter 4

Micro-machining

The heat produced by electrochemical discharges can be used to locally attack a substrate. This can be done by localized melting of the material or by thermal promotion of chemical etching [3, 4]. Combinations of both are possible as well. This technique, first introduced in Japan by Kurafuji and Suda [73], is called electrochemical discharge machining (ECDM) [74], Electrochemical Spark Machining (ECSM) [75] or Spark Assisted Chemical Engraving (SACE) [76].

Glass micro-machining was the first application of SACE. Other non-conducting materials can also be machined using this technology (e.g., granite, refractory fire-brick, aluminum oxide, plexiglas, quartz) [77–81]. Concretely, an electrode (either a cathode or an anode) has to be brought in close vicinity of the work-piece (typically closer than 25 μm in the case of glass [82]). In the context of micro-machining, the electrode surrounded with electrochemical discharges is called the tool-electrode and the electrochemical cell terminal voltage the machining voltage. The attack being localized, controlled structures can be machined. Typically holes from 100μm to 1 mm diameter, and micro-channels of a few hundred micro-meters width/depth can easily be machined in glass. Applications are mainly to be expected in the field of micro-fluidics and lab-on-a-chip devices.

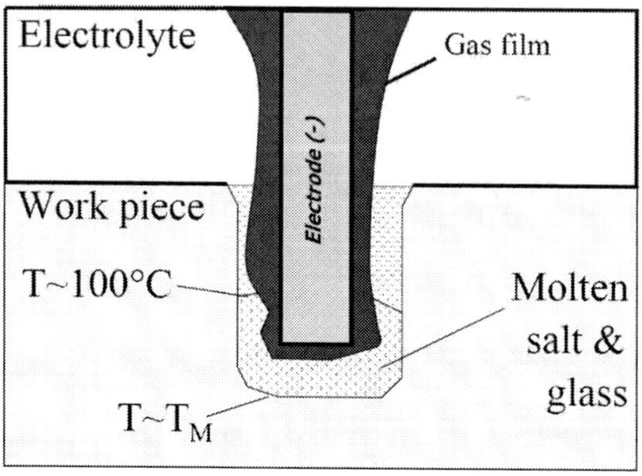

Figure 4.1. Schematic of glass machining with electrochemical discharges. T_M is the temperature needed in order machining takes place. Reprinted from [3] with permission from Elsevier.

4.1. Material Removal Mechanism

Material removal mechanism in machining with electrochemical discharges is a complex process and investigations are still underway. Several effects may contribute to the material removal [74, 83], including:

- melting and vaporization due to electrochemical discharges,
- high-temperature chemical etching,
- differential expansion of constituents and weathering,
- random thermal stresses and micro-cracking and spalling, and
- mechanical shock due to expanding gases and electrolyte movement.

So far only thermal effects and chemical etching have been investigated systematically, mostly for glass machined with a cathode-tool. The following discussion will therefore concentrate on glass micro-machining.

In the case of glass machining using an alkaline electrolyte, the currently admitted material removal process is depicted on figure 4.1 [3]. The electrochemical discharges heat up locally the work-piece, which results in lowering of the viscosity of glass and enhancing of chemical etching by OH radicals from the electrolyte. A zone of glass with lowered viscosity and molten salt forms in front of the tool-electrode (water evaporates at that temperature). If sodium hydroxide is used as the electrolyte, etching takes place probably by complexation of silicate as will be discussed in section 4.1.2.. The temperature needed in order material removal takes place is called the machining temperature T_M. Experimental and numerical simulations indicates that this temperature is about $500-600^\circ\,C$ [84, 85].

4.1.1. Thermal Effects

As the electrochemical discharges are the heat source, machining at low and high depths will be very different. With increasing depth, difficulty for the electrolyte to flow towards the tool-electrode tip will increase (especially in the case of micro-hole drilling). Consequently, discharge activity and chemical etching will be reduced. Both effects result in a significant lowering of the material removal rate.

At low depth machining, defined as the situation where the electrolyte can reach the machining zone without difficulty, the material removal rate is essentially limited by the time needed to reach the machining temperature, i.e. characterized by the work-piece's thermal diffusivity and conductivity λ. A simplified model, considering the propagation of heat inside the work-piece, allows the estimation of the material removal rate. For glass the following rate is predicted [3, 85]:

$$\frac{dz}{dt} = \frac{4a}{b}\frac{\kappa-1}{2\kappa}\left[1-\exp\left(-\frac{\pi(\kappa-1)^2}{\kappa^2}\right)\right]. \tag{4.1}$$

Here z is the depth of the machining in function of time, b the tool-electrode radius and κ is the ratio between the mean heat power transferred to the work-piece by the electrochemical discharges and the minimal heat power $p^{min} = \lambda\pi(T_M - T_0)$ needed in order machining can take place (the machining temperature can be reached). The ratio κ is directly proportional to the voltage applied during machining [85]. Machining at low depth is termed to be in the discharge regime, because material removal rate is essentially controlled by the discharge activity [86].

High depth machining is defined as the situation where the electrolyte cannot reach the machining zone easily. As a consequence, material removal by chemical etching or melting of the work-piece is lowered. The process is limited by the ability to bring the electrolyte to the machining zone and to remove the machined material out of it, both of which are geometry dependent aspects. As local flows controls material removal rate, machining in this case is termed to be in the hydrodynamic regime [86].

4.1.2. Chemical Effects

In micro-machining using electrochemical discharges, the effect of the electrolyte on machining is complex and cannot be described solely as a function of its concentration and temperature. The nature of the electrolyte strongly influences the machining behavior. In general, acid electrolytes result in very poor material removal rates compared with alkaline ones [87]. The surface roughness of the machined work-piece is influenced by the electrolyte. In particular, molten salt electrolytes (e.g., a eutectic of NaOH and KOH melting at 170° *C*) can drastically improve the smoothness of the machined surface [78]. The most common used electrolytes for glass machining are NaOH and KOH.

In the case of glass machining with sodium hydroxide, the work-piece is probably etched [16, 74, 82, 87–89]:

$$2NaOH + SiO_2 \rightarrow Na_2SiO_3 + H_2O. \tag{4.2}$$

The sodium silicate formed is removed by the electrolyte. This chemical reaction is strongly enhanced by increasing the temperature [16, 87]. Similar results are reported for ceramic materials [89] and quartz [90]. Yang et al. [87] defined the process as a high-temperature etching. By comparing the surface quality obtained at high-temperature etching of glass with NaOH, electrical discharge machining of indium-tin oxide (ITO), and glass machining by electrochemical discharges, they concluded that SACE is mostly like to be a combination of thermal effects and chemical etching.

Energy dispersive X-ray (EDX) analysis of the machined structures for sodalime glass confirms that Na^+ concentration decreases whereas H^+ concentration increases in the machining area [16] suggesting that besides etching, leaching occurs as well. Local surface modification may also occur depending on the electrolyte used. Basak [32] suggested that in the case of KOH, the larger K^+ ions from the electrolyte replace Na^+ ions of the glass surface and decreases its strength. This effect may explain why higher material removal rates are observed for KOH compared with NaOH for the same concentrations and machining voltages [87].

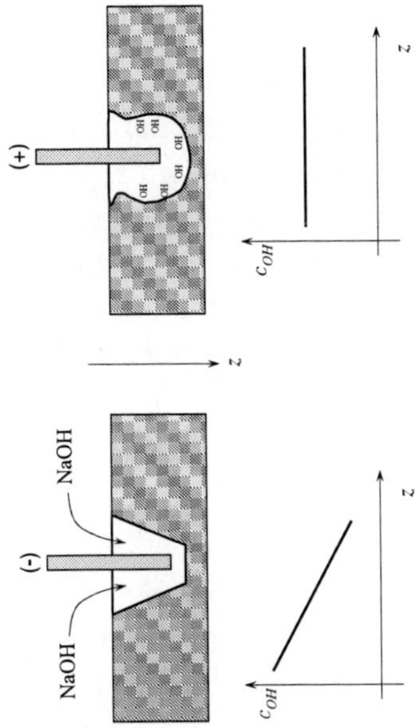

Figure 4.2. Comparison of machining with a cathode- and an anode-tool. In the case of a cathode, the concentration of OH radicals decreases with the hole depth resulting in a conical shape, whereas in the case of an anode, the Hickling and Ingram mechanism will produce an uniform OH concentration along the hole resulting in a spherical shape. Reprinted from [3] with permission from Elsevier.

Recently, Didar et al. [91] showed that micromachining with electrochemical discharges changes the mechanical properties of glass surface. Using nano-indentation tests, the authors showed that the hardness and density of glass increases. This effect is attributed to the rapid cooling that takes place right after machining.

The polarity of the tool-electrode also influences greatly the machining process [78, 90]. As shown first by West and Jadhav [92], drilling with a anode-tool results in more spherical hole shapes than with cathodic machining. This effect may be attributed to the formation of a large number of OH radicals at the gas film surface by the Hickling and Ingram mechanism during anodic electrochemical discharges [3]. The concentration of OH radicals, as a function of hole depth z, is more or less uniform. The work-piece is etched isotropically resulting in a spherical shape (Figure 4.2). This is not the case when using a cathode as the tool-electrode. In this case, only OH radicals from the electrolyte (NaOH) are present. During drilling, electrolyte is mostly present at the top of the hole. The concentration of OH radicals decreases with z, resulting in a larger entrance diameter. This mechanism will result in a conical hole shape. As more OH radicals are present in the case of an active anode than for an active cathode, chemical etching is also more important. Consequently, the surfaces are smoother than those obtained by cathodic machining [90, 92]. However, when using anodic polarization, the tool-electrode will be anodically dissolved resulting in high tool wear.

4.2. Micro-hole Drilling

To achieve material removal, the heat source produced by the electrochemical discharges has to be in close vicinity to the work-piece (typically closer than 25 μm [82] in case of glass). This is achieved with various feeding mechanisms of the tool-electrode.

The simplest feeding mechanism is gravity-feed in which a constant force is applied to the tool-electrode in order to guarantee mechanical contact with the work-piece. A second option is to move the tool-electrode at a speed lower than the mean material removal rate. Finally, a third way is to control the tool feed as a function of the status of the actual machining process. Although this last strategy is highly desirable, it has never been achieved to date. Preliminary, very promising results were published by Morrison et al. [93] recently.

4.2.1. Gravity-feed Drilling

Gravity-feed is a very popular strategy for drilling because of its simplicity and ability to give excellent results in term of material removal rate and quality of machined holes. As during gravity-feed drilling the tool-electrode is constantly in contact with the work-piece, drilling progress can be monitored by measuring the motion of the tool. Tool-electrode wear will affect this measure. However, the wear of SACE tool-electrodes is generally quite low and remains usually in the range of a few microns, even when drilling over several minutes [73]. Typical examples for glass drilling, with a cylindrical stainless steel cathode-tool in 30 *wt*. NaOH, as a function of various machining voltages are shown in figure 4.3. After a first phase (discharge regime), where the drilling speed is high, a progressive slow down of the material removal rate is observed until

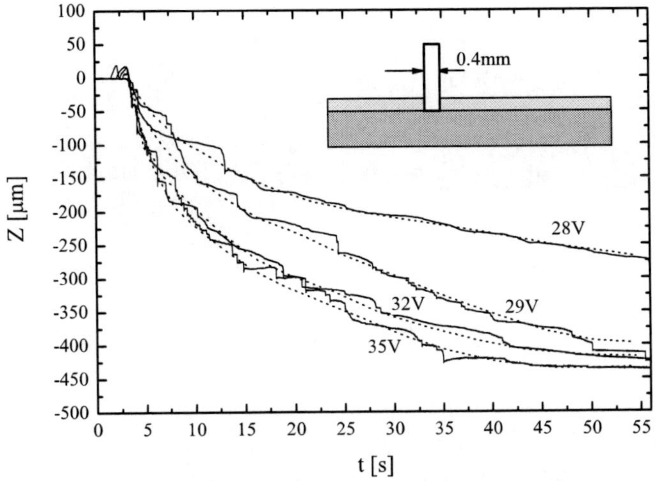

Figure 4.3. Typical evolution at various voltages of SACE glass gravity-feed drilling using a cylindrical tool (cathode) of 0.4 *mm* diameter with a force of 0.8 *N* acting on it. The electrolyte (30 *wt.%* NaOH) level above the workpiece is about 1 *mm*. Reprinted from [86] with permission from IOP Publishing.

a limiting drilling speed v_{lim} almost independent of the machining voltage is reached (hydrodynamic regime). The existence of the two machining regimes results in different qualities of micro-holes as a function of the machining voltage and depth. The drilled micro-holes can be classified into four types (Figures 4.4 and 4.5) [94]:

1. Well defined cylindrical contours with smooth surface (Figure 4.4(a)): characteristic for low depths (100 μm, $28-37$ V) or low machining voltages (28 V, up to 300 μm).

2. Jagged outline contours (Figure 4.4(b)): appear at depths between 200 and 300 μm using a machining voltage of about 30 V.

3. Hole with heat affected zone (Figure 4.4(c)): obtained for machining voltages above 30 V and depths higher than 100 μm. The contour remains cylindrical.

4. Hole with thermal cracks (Figure 4.4(d)): when machining at 37 V and higher, the hole is characterized by cracks and a large heat affected zone.

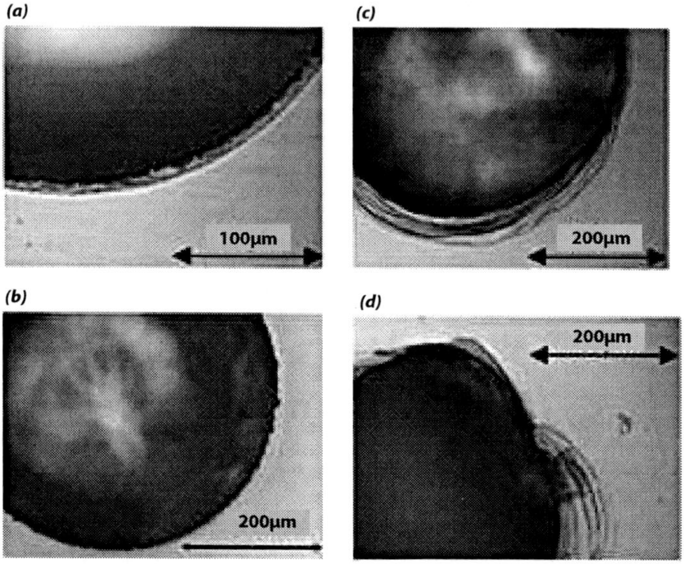

Figure 4.4. Different micro-hole qualities obtained by gravity-feed drilling with a stainless steel tool-electrode of 0.4 *mm* diameter in glass. Reprinted from [94] with permission from IOP Publishing.

The entrance holes are generally no longer cylindrical.

These various types of micro-holes, as well as the evolution of the mean micro-hole diameter in function of depth and voltage (Figure 4.6), can be explained in function of the two machining regimes. Therefore follow the drilling at a given voltage [94]. Initially, machining is done solely in the discharge regime (down to about 100 μm). The quality of the hole is not affected by the machining voltage. A higher voltage results in faster drilling without loss of quality and increase of hole diameter (zone A in figure 4.6). As the tool drills deeper into the work-piece, the difficulty in removing the machined material and insufficient wetting of the tool tip increases. The machining moves into the hydrodynamic regime. More and more discharges take place in the upper part of the tool, increasing the diameter proportionally to the machining voltage (zones B and C in figure 4.6). Jagged contours appear. As the drilling continues, the hole diameter increases further until it reaches its maximal value (depending on the machining time and voltage). In this situation, the border of the hole is not

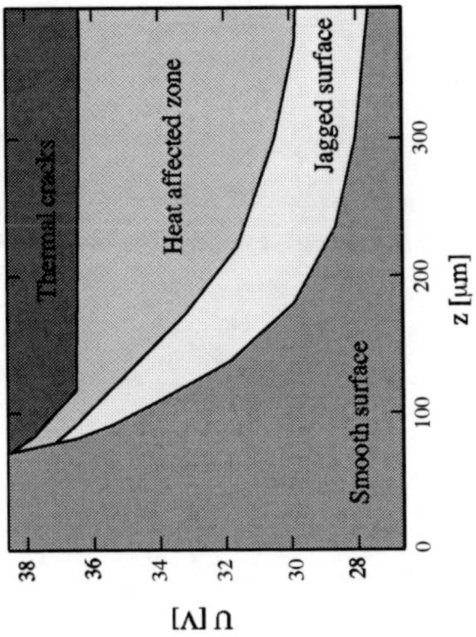

Figure 4.5. Evolution of SACE glass gravity-feed drilling in the machining voltage-drilling depth plane. Reprinted from [94] with permission from IOP Publishing.

heated enough to be etched. The micro-hole does not increase any longer but a heat affected zone starts growing instead. If the heat power supplied is too high (typically for voltages above 36 *V*), thermal cracks appear.

The dimensions of the micro-holes are as well affected by the electrolyte composition and bulk temperature. The mean micro-hole diameter increases with lower electrolyte concentration [87]. This effect is attributed to the reduced machining speed at low electrolyte concentration, resulting in significantly higher drilling times. The chemical composition of the electrolyte has even more dramatic effects. For example, the use of KOH results in almost half as large diameters compared with the holes drilled using NaOH [87]. This is due to the different drilling times resulting from the use of KOH and NaOH electrolytes.

Another important aspect is the achieved surface roughness of the drilled holes. Typical values for glass can be as low as a few hundred nanometers [87]. The use of NaOH results in smoother surfaces than KOH [87]. This is probably due to the lower heat available for machining in NaOH compared with KOH at a given machining voltage (due to the different electrical conductivities of NaOH and KOH).

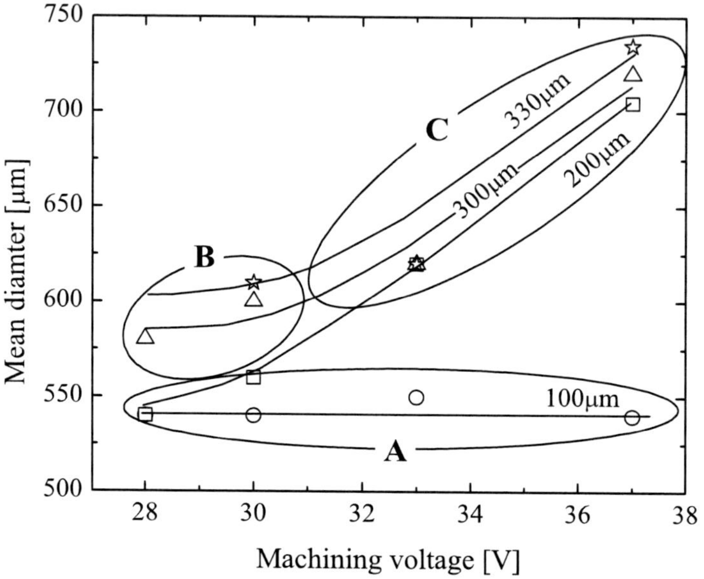

Figure 4.6. Mean entrance diameter of micro-holes obtained by SACE glass gravity-feed drilling as a function of the machining voltage and the drilling depth for a 0.4 *mm* stainless steel tool-cathode. Reprinted from [94] with permission from IOP Publishing.

4.2.2. Constant Feed Drilling

Another popular machining strategy for SACE is constant velocity feed drilling. Compared with gravity-feed drilling, mechanical contact between the tool-electrode and the work-piece is avoided. This can overcome problems like tool-electrode bending. However, as the tool-feed rate is selected beforehand, the gap between the work-piece and the tool-electrode is not controlled and is generally not constant during machining. One of the consequences is that the progress of drilling cannot be monitored online.

Further, the tool-electrode feeding rate has to be selected properly. Feed rates higher than the mean material removal rate of the process will result in breaking either the work-piece or the tool-electrode. On the other hand, very slow feed rates will increase drilling times and may result in large heat affected zones around the micro-hole. Typical values for tool feed rates reported in the literature range, depending on the tool-electrode diameter, from 5 to 15 $\mu m \cdot$

s^{-1} [40, 80, 95], a value slightly higher than the limiting speed reached in the hydrodynamic regime during gravity-feed drilling.

As in gravity-feed drilling, the mean hole diameter increases with drilling depth [80] until a maximal diameter is reached. The feeding rate directly affects the micro-hole surface roughness. As shown by Han et al. [95], the surface roughness follows an inverse volcano dependence of the feeding rate. This dependence is due to the competition between local heating and chemical high-temperature etching of the glass.

The main advantage of constant velocity feed drilling compared with gravity-feed drilling is that the motion of the tool-electrode is user controlled. This allows the machining of more complex shapes than cylindrical holes. An impressive example is the machining of threads in glass as shown by Lee et al. [96].

4.3. 2D and 3D micro-machining

In 2D SACE machining, the tool-electrode is moved above the glass surface (closer than $20-25\ \mu m$) by following a specified path at a constant travel speed. Typical values are around $50-150\ \mu m \cdot s^{-1}$. Such high values are possible as, in contrast to drilling, the machining is almost never limited by the ability of the electrolyte to reach the machining zone. The combination tool travel speed and machining voltage has to be selected appropriately. For example, when machining at 28 V it is not possible to have any acceptable channel shape for travel speeds higher than $40\ \mu m \cdot s^{-1}$. Increasing the machining voltage will allow to increase the tool travel speed. On the other hand, the utilization of a too low tool travel speed results in a non-smooth channel surface with significant depth variation along the channel. There are optimal conditions under which micro-channels with edges and surfaces of acceptable quality are achieved. The channels so obtained can be categorized as follows (Figure 4.7) [97]:

1. Well-defined linear channel edges and smooth channel surface: obtained with low machining voltage (less than 32 V) and appropriate tool travel speed; e.g. 28 V together with the tool travel speed ranging from 5 to $10\ \mu m \cdot s^{-1}$ (Figure 4.7(a)).

2. Jagged outline contours with smooth channel surface: obtained with low machining voltage (less than 32 V) and tool travel speed lower than in

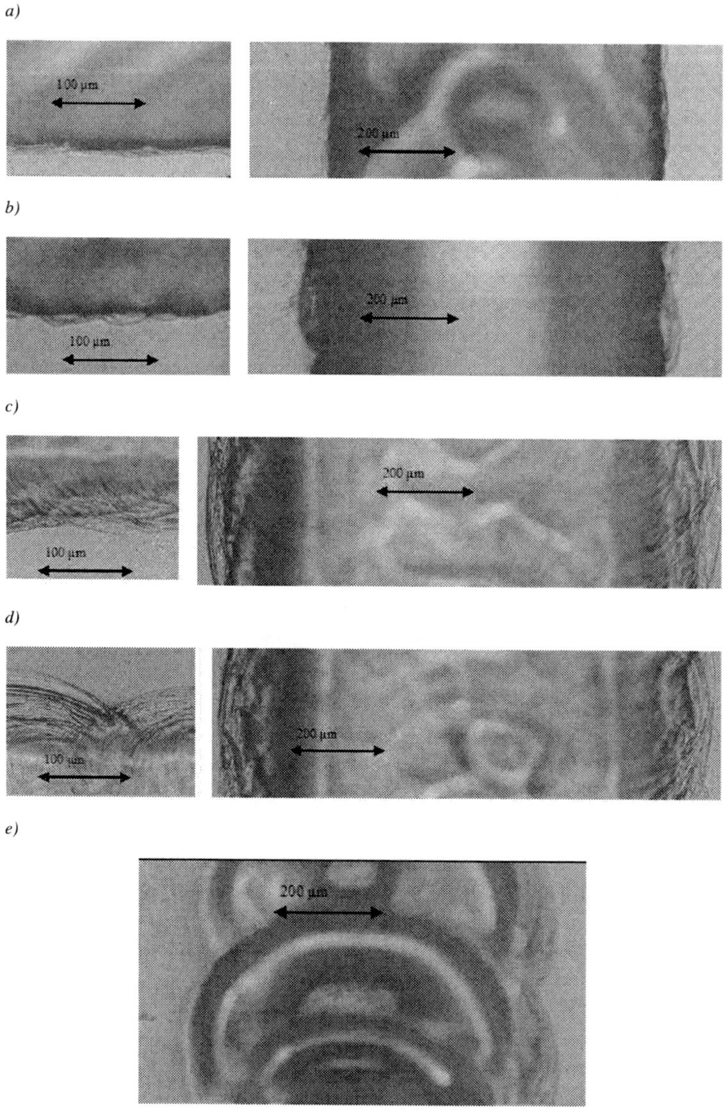

Figure 4.7. Different types of channels in 2D-SACE machining. Machined with a 0.2 *mm* cylindrical stainless steel tool-cathode in 30 *wt.%* NaOH. Reprinted from [97] with permission from IOP Publishing.

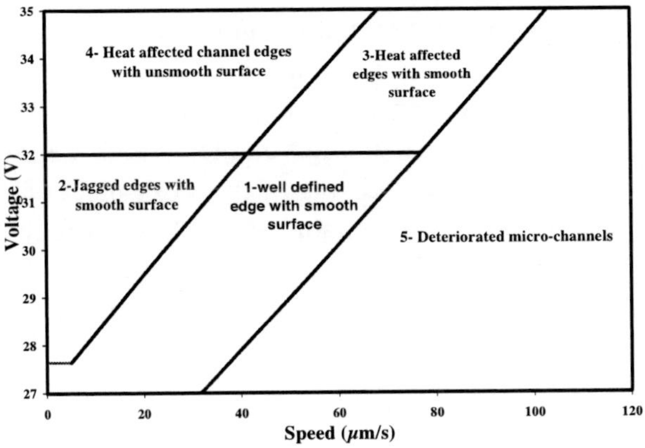

Figure 4.8. Characterization diagram for micro-channels machined using SACE technology as a function of the machining voltage and the tool travel speed (for tool-electrode distances less than 15 μm). Machined using a 0.5 mm stainless steel cathode in 20 wt.% NaOH. Reprinted from [97] with permission from IOP Publishing.

the previous contour type. The micro-channel edges are jagged but the channel surface is still flat and smooth (Figure 4.7(b)).

3. Heat affected edges with smooth channel surface: obtained with high machining voltage (more than 32 V) and a tool travel speed high enough to remove the machined material. The micro-channels exhibit a smooth channel surface but with not well defined channel borders (Figure 4.7(c)).

4. Heat affected edges with non-smooth channel surface and thermal cracks: obtained with high machining voltage and low tool travel speed (e.g. 32V with speed less than 30 $\mu m \cdot s^{-1}$). The edges are unclear and heat affected. The surface is neither flat nor smooth. Thermal cracks are observed (Figure 4.7(d)).

5. Deteriorated micro-channels: obtained for too high tool travel speed. The channels are no longer continuous and the inner surface becomes very rough. For example, at 28 V increasing the tool travel speed above 40 $\mu m \cdot s^{-1}$ results in deteriorated micro-channels (Figure 4.7 (e))

Figure 4.8 summarizes the type of micro-channel obtained for different combinations of machining voltage and tool travel speed. Machining at less than 32 V

result in acceptable micro-channel quality and an appropriate selection of the tool travel speed will result in the best quality.

The depth of the channel cannot be monitored directly and is function of the tool travel speed and the machining voltage. Another important aspect to take into account is the chemical etching of the work-piece. Due to this effect, the depth of the channels will not remain constant over machining time, but on the contrary increases slightly [97]. Two parameters are used to characterize the depth z of a machined micro-channel in function of machining time: the channel depth z_0 at time zero and the average channel depth increase rate m_d:

$$z(t) = m_d t + z_0. \tag{4.3}$$

Both parameters are functions of the machining voltage, as it controls the heat transferred by the electrochemical discharges to the work-piece and the tool travel speed [97].

4.4. Controlling the Machining Process

The electrochemical discharges heat up the machining zone locally resulting in softening and melting of the work-piece. Material is removed by high-temperature chemical etching. Therefore, controlling the machining process means taking control of the following aspects:

- stable discharge activity,
- localized heating,
- and high-temperature chemical etching.

Optimal surface quality is achieved by maximizing the chemical etching and at the same time minimizing the local heating in order to avoid the formation of heat affected zones and thermal cracks:

Promoting chemical etching : Chemical etching can be promoted by optimizing the local electrolyte supply to the machining zone. This aspect is particularly important for micro-drilling where with increasing hole depth the flow of the electrolyte to the machining zone becomes more and more difficult. The following strategies were explored so far: optimizing the tool-electrode shape, promotion of the electrolyte flow by tool-electrode motions (e.g. tool vibrations [92, 96, 98, 99] and tool rotation

[87, 96, 100, 101]), electrolyte motion [36] and flushing of the micro-hole in function of the measured force exerted on the tool-electrode [99, 102].

Optimizing heat transferred to the work-piece : Controlling the heat generated can be achieved passively or actively. Passive methods try to keep the heat generated in the machining zone within a reasonable range by removing excess energy through heat transfer. Heat transfer can be influenced by acting on either the heat conduction of the electrolyte [95, 103] or of the tool-electrode [72]. Active methods try to keep the heat supplied to the work-piece in a well defined range by controlling the amount of energy supplied to the machining zone. This is achieved either by using small machining voltages [28] or by reducing the time during which the voltage is applied [101, 102, 104, 105]. Another promising approach is to control the discharge location by using side insulated tool-electrodes [36, 106].

Among the various strategies proposed so far, the utilization of pulsed voltage supply is among the most promising [101, 102, 104, 105]. Two major benefits can be achieved. First, one can get rid of the problems associated with the instability of the gas film by periodically destroying and rebuilding the gas film. At the same time one promotes the flow of the electrolyte towards the machining zone. Second, by choosing the ratio between the pulse-on time and the pulse-off time, one can optimize the heat transferred to the work-piece. The pulse-on time has to be selected longer than the mean gas film formation time, but short enough to avoid excessive heat transfer to the work-piece. With the pulse-off time one can allow the system some time to cool down and thus avoid problems like jagged contours and heat affected zones. The major drawback of pulsed voltage machining is the reduced material removal rate. The use of pulsed voltages for micro-machining of glass with electrochemical discharges was first proposed by Kim et al. [104] and Zheng et al. [101, 105]. They showed that excellent results can be achieved for micro-hole drilling and 3D micro-structuring, especially in combination with tool rotation (see figure 4.9).

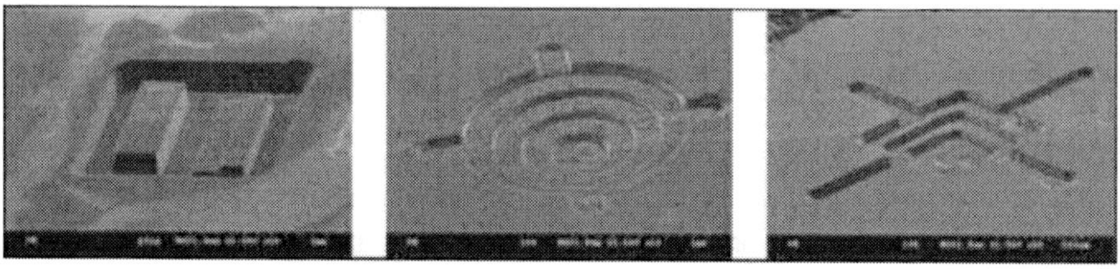

Figure 4.9. Examples of 3D microstructures in Pyrex glass machined using electrochemical discharges. Reprinted from [105] with permission from IOP Publishing.

Chapter 5

Nanoparticle Synthesis

At the date of today, the synthesis of ultra-fine particles is the most recent application of the electrochemical discharge phenomenon in aqueous electrolytic solutions. An overview of the literature about the fabrication of nanoparticles by this template technique is briefly reported in this chapter emphasizing the used experimental and characterization methods. Discharge electrolysis in molten chloride in argon atmosphere to synthesize silver and titanium particles proposed by Kawamura et al. [107] and Oishi et al. [108] is discussed as well. In the last section, a mechanism of nanoparticles synthesis by electrochemical discharges is proposed where the nanoparticles nucleate from the reduction of metal ions by high energy species and electron-ion recombination followed by coagulation-agglomerations as the growth process. The erosion of the electrode and the thermal disintegration of its material are not excluded to explain this nanometric application of the electrochemical discharges.

5.1. Nanoparticles of Ni and Cu

5.1.1. Experimental

Recent investigations have been conducted by the authors to synthesize colloidal nanoparticles of nickel and copper using the electrochemical discharge phenomenon [109]. The electrochemical cell consists of two electrodes where the anode is of much larger surface then the cathode. As an example, the authors have chosen a nickel rod of 0.5 mm diameter immersed into a 2 mol/l sulphuric acid solution such as its surface is 5 mm^2. The anode is a 20 cm^2

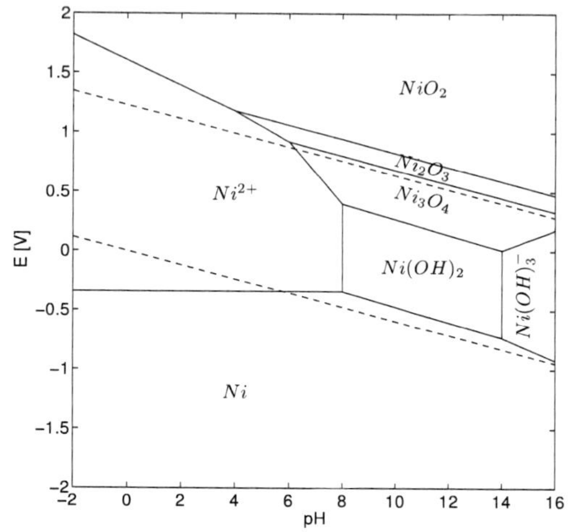

Figure 5.1. Potential-pH diagram of Ni $10^{-4\ M}$ at 25 °C.

nickel foil. The same experimental conditions hold for copper nanoparticles synthesis where both electrodes were made of copper. Prior to the experiments, rapid metal decontamination in a three to one solution of hydrochloric acid and nitric acid is performed for both electrodes followed by subsequent abundant washing with deionized water. The pH of the solution was chosen strongly acid to obtain stable ionic entities generated from the anodic side. For example, figure 5.1 illustrates a simplified version of the Pourbaix diagram of nickel at 10^{-4} $mol \cdot l^{-1}$ concentration at an ambient temperature of 25 °C (not all possible reactions are taken into account). Pourbaix diagrams allow, for a given metal, to specify at which region in the potential-pH space to operate to reduce the metallic cations. This justifies the choice of the acid environment for the electrochemical solution.

As far as it concerns the electrical conditions used for nanoparticle synthesis, the steady state current-voltage characteristics dictates the required potential difference to obtain colloidal nanoparticles by electrochemical discharges. As emphasized earlier, the terminal voltage has to be above the critical voltage in order to allow electrochemical discharges to occur. In the experiments the potential difference was 32 V and stand for 3 hours. It was observed that the solution color turned from transparent to light green to olive dark green.

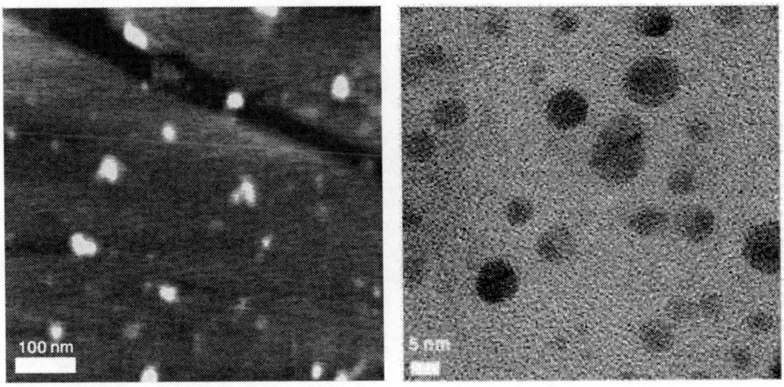

(a) AFM (b) TEM nanograph

Figure 5.2. Nickel nanoparticles synthesized by electrochemical discharges: (a) AFM and (b) TEM nanographs. Reprinted from [110] with permission from Elsevier.

5.1.2. Characterization

Nanographs of nickel nanoparticles are illustrated in figure 5.2. The two-dimensional topographic measurements with AFM (Atomic Force Microscopy) were performed using a freshly cleaved muscovite Mica substrate where the colloidal solution of nanoparticles was deposited and dried. The particles size ranges from about 50 *nm* down to a few nanometers, according to the high magnification of the TEM (Transmission Electron Microscopy) nanograph in figure 5.2(b).

Copper nanoparticles were successfully synthesized by electrochemical discharge phenomenon, as illustrated in figure 5.3. Particle size ranges from around 30 down to 5 *nm* where the agglomeration is more notable than for nickel. The EDX (Energy Dispersive X-Ray Spectroscopy) spectrum suggests that the copper particles are oxidized.

Following to the synthesis procedure of nanoparticles, systematic characterization of the products is done with cyclic voltammetry. Therefore the nanoparticles were fixed on carbon. 10 *mg* of carbon black Vulcan XC-72 were added to the solution followed by overnight stirring. The nickel-coated carbon black colloidal solution was filtered, dried and dispersed in an ultrasonic bath. The working electrode for cyclic voltammetry measurements were fabricated by deposing 5 μl of the Ni/C ink in addition to 5 μl of 20 times diluted Nafion

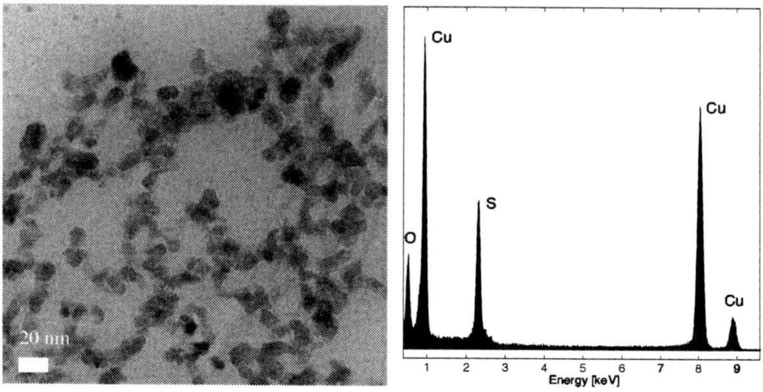

(a) TEM nanograph (b) EDX spectrum

Figure 5.3. TEM nanograph (a) with its EDX spectrum (b) of copper nanoparticles synthesized by electrochemical discharges. Reprinted from [110] with permission from Elsevier.

10 $wt.\%$ on a carbon graphite electrode. Without getting much into the detailed reactional electrochemistry of the nickel element [109], figure 5.4 shows the reduction and oxidation peaks, typical for nickel behavior, in a 1 $mol \cdot l^{-1}$ of potassium hydroxide solution at room temperature with a 20 $mV \cdot s^{-1}$ scan rate versus the Ag/AgCl reference electrode. The counter electrode used in the three-electrode setup is an ultra high purity platinum foil. In this example the repeatability of the process is verified as well, through the correspondence of the voltamogramms of four different experiments realized under the same conditions.

5.2. Nanoballs of Ni, Ti, Ag and Au

Toriyabe et al. [111] reported the controlled formation of metallic nano-sized spheres of several types of transition metals i.e. Ni, Ti, Ag and Au, by generating discharge plasma in electrolytic medium. Low concentrated potassium bicarbonate was used as the electrolytic solution. The sacrificial electrodes, cathodically polarized, were thin wires of the several tested metal elements. The anode was a platinum mesh of much larger surface than the cathode in order to displace the voltage breakdown to the cathodic side of the electrochemical cell. The electrical conditions were several times higher than the ones in previous

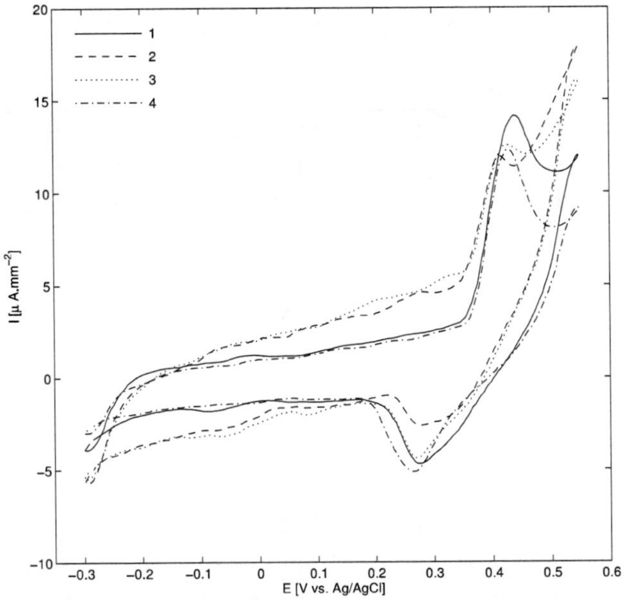

Figure 5.4. Cyclic voltamogramms in 1 M KOH of four samples from four different experiments of carbon supported Ni nanoparticles synthesized by electrochemical discharges. Scan rate 20 $mV \cdot s^{-1}$.

cited works due to the low electrical conductivity of the K_2CO_3 solution as well as the ratio between the electro-active surfaces of the cathode and of the anode.

According to SEM (scanning electron microscope) micrographs of the cathodes' surfaces, relatively monodispersed nanospheres of Ni, Ti, Ag and Au of a few hundreds of nanometers can be observed and witness of the versatility of this route for nanoparticles fabrication. TEM nanographs and EDX diffraction patterns confirm the chemical composition of the colloidal nanoparticles collected from the solutions. By varying the applied terminal voltage, they have reported that the nanoparticles size is inversely proportional to the potential difference and this was valid for the different metals they investigated. The shape of the products is correlated to the applied voltage as well.

5.3. Nanoparticles of Cu, Pt, Au and Pt/Au

In [5] is reported the successful synthesis of nanoparticles of copper, platinum and gold using platinum-made electrodes immersed into solutions containing precursors of the metals. Low concentration salts in the range of a few *mM* were diluted and play the role of the cationic source. For example, Cu nanoparticles were synthesized in $CuSO_4 + H_2SO_4$ while the Pt nanoparticles were obtained in $H_2PtCl_6 + HClO_4$ solution.

Lal et al. [5], in addition to the quasi non-convective situation of the nanoparticle synthesis, added Corriolis forces to the process by using a rotating disc electrode turning up to 5000 *rpm*. They generated low strength potential vortices around the working electrode such that the fabricated particles are ejected from the electron-cation recombination region. This is a way for the particle size control and also to prevent the particles from deposition at the working electrode surface. From the Boltzmann equation, the computation of the ratio between the probabilities of finding a particle of a given diameter to not finding any particle in a control volume within the cationic reduction zone shows correspondence with the experimental results. According to the calculations, the clusters size ranges from 190 *nm* down to slightly above 60 *nm* with cathode rotation speed rising from 250 to 5000 *rpm* where experimentally the group has found an interval ranging from 140 down to 80 *nm* particle diameter.

5.4. Fine Particles Formation in Eutectic Melts

By generating quasi-stable cathodic discharges in eutectic melts under atmospheric pressure of argon, Kawamura et al. [107] and Oishi et al. [108] successfully synthesized fine metallic particles. Experimentally, the group used a sub-kV DC power supply and a two-electrode configuration cell under atmospheric argon flow. The sacrificial anode, fully immersed into the eutectic melt, was made of the metal type of the targeted nanoparticles. The molten salt was composed of 58.5 *mol %* of LiCl and 41.5 *mol %* of KCl heated up to a few hundreds of degrees Celsius. Additional low concentrated metal salt was added to the melt i.e. AgCl in the case of silver nanocrystals or K_2TiF_6 in the case of titanium nanoparticles synthesis. The electrical discharges were produced by carefully approaching the tungsten cathode tip to the molten salt surface and removing it a few millimeters back. The discharge is stationary for certain time duration, so that the same procedure had to be repeated again once the electronic activity at the cathode tip ceases.

Several characterizations have been carried out, namely SEM (Scanning Electron Microscopy) micrographs for nanoparticles morphology observation and XRD (X-ray Diffraction) patterns, XPS (X-ray Photoelectron Spectroscopy) and EPMA (Electron Probe micro-Analyzer) for chemical and crystallographic structure. Fine clusters with size ranging from a few hundreds of nanometers down to 10 *nm* were obtained by cathode discharge electrolysis in molten salts. The group of research has suggested some correlations controlling the particles sizes and shapes by analyzing the discharges behavior, the time of the discharging, the distance between the cathode tip and the bath surface as well as the chemical composition of the salt.

5.5. Mechanism of Nanoparticle Synthesis

Two categories of mechanisms for the nanoparticle synthesis are proposed in this section. Each phase in the electrochemical discharge system i.e. *i)* the gas-liquid interface, *ii)* the solid phase represented by the electrode surface and *iii)* the gaseous phase represented by the gas film are believed to play some roles in the obtention of the fine particles. Naturally, the presented mechanisms can occur simultaneously, but at different time scales and with different yields.

Firstly, a mechanism based on the metal ion reduction down to the zero-valent state is discussed. Mainly adapted from radiative chemistry, this model occurs at the gas film-electrolyte interface. The possible reactions of reduction and oxidation by the solvated electron and H and OH radicals, considering this mechanism, are presented. The second category of mechanisms considers the physics of the phenomenon where of the electrode sputtering by ion bombarding and thermal melting of the electrode material are considered. To a certain extent, the electrochemical discharges can be considered, in the same time, as a bottom-up and up-down routes to synthesize the fine clusters. In other words the nucleation of atomic particles, at zero-valent state, will be self-assembled depending on various parameters such as the electrolyte temperature, pressure, viscosity, concentration, nature of the metal etc.

5.5.1. Cation Reduction Mechanism

The first mechanism, proposed to explain the nanoparticle synthesis by electrochemical discharge phenomenon, is adopted from high energy radiation chemistry of aqueous solutions. Is is proposed that this is the dominating mechanisms in the experiments discussed in previous sections. As the gas envelope covering the cathode constitutes an insulating gaseous body, the discharges from the cathode toward the electrolyte through the gas film are somehow similar to radiative beams generated from a particle beam at the top of a solution. A similar assumption was already made by Hickling and Ingram for the anodic glow discharge electrolysis mechanism, where the charges were carried out by the superoxide ion H_2O^+.

The efficiency of the reduction of metal ions M with valency $z+$ by the solvated electron e_s^- is very high (see chapter 3) and shall follow the scheme from reaction 5.1 to reactions 5.4 [110]:

$$H_2O \rightsquigarrow e_s^-, H^\cdot, OH^\cdot, H_2, H_2O_2, H_3O^+ \quad (5.1)$$

$$e_s^- + M^{z+} \longrightarrow M^{(z-1)+} \quad (5.2)$$

$$e_s^- + M^+ \longrightarrow M^0 \quad (5.3)$$

$$n\,M^0 \longrightarrow M_2 \longrightarrow \cdots M_n \cdots \longrightarrow M_{agg} \quad (5.4)$$

Similarly, with the $H^\cdot$ atoms, the reduction of metal ions can be written:

$$H_2O \rightsquigarrow e_s^-, H^\cdot, OH^\cdot, H_2, H_2O_2, H_3O^+ \quad (5.5)$$

$$H^\cdot + M^{z+} \longrightarrow M^{(z-1)+} + H^+ \quad (5.6)$$

$$H^\cdot + M^+ \longrightarrow M^0 + H^+ \quad (5.7)$$

$$n\,M^0 \longrightarrow M_2 \longrightarrow \cdots M_n \cdots \longrightarrow M_{agg} \quad (5.8)$$

In the other hand, the $OH^\cdot$ radical is the strongest oxidizing agent, with $E°(OH^\cdot/OH^-) = 1.90\ V$ in neutral solution and $E°(H^+, OH^\cdot/H_2O) = 2.72\ V$ in acidic solution. At room temperature, both e_{aq}^- and $H^\cdot$ atom are able to reduce quite easily metal ions to the zero-state of valence. In contrast, the $OH^\cdot$ radicals oxidize the atoms or ions present in the solution to a higher level of oxidation. Reaction 5.4 or 5.8 where n M^0 aggregate progressively to the final stable state M_{agg} can be interrupted and reversed by oxidation at any stage. Oxidation reaction by either the hydronium ion H_3O^+ or $OH^\cdot$ radical has been demonstrated [112]. This reaction can be prevented by adding scavenger species of $OH^\cdot$ radicals, such as secondary alcohols or formate anions. The secondary formed radicals are strong reducing species and take over the reduction reaction instead of the oxidation reaction by the $OH^\cdot$ radical [113]. It is understood that oxidation reactions by the $OH^\cdot$ radicals happen in aqueous solutions. In summary, this

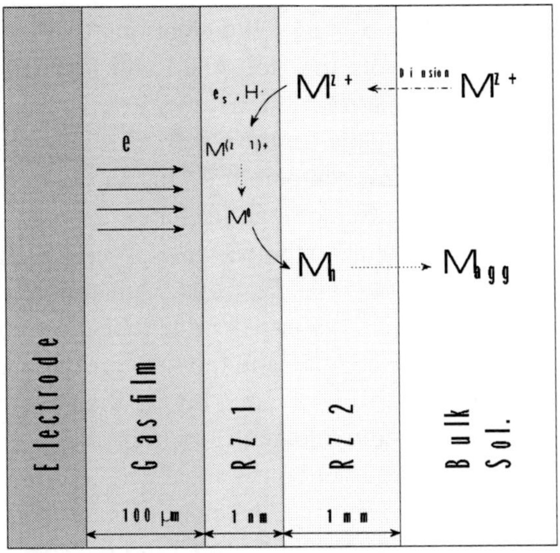

Figure 5.5. Primary and secondary reaction zones. Reprinted from [110] with permission from Elsevier.

so-called gas-liquid interface mechanism is based on results adapted from high linear energy transfer radiation chemistry of aqueous solutions. The reduction of the metal cations M occurs in the primary reaction zone of Hickling and Ingram, which was proposed initially for the anodic GDE phenomenon. An adapted version of the Hickling and Ingram mechanism, with the two reaction zones, to the electrochemical discharges is illustrated in Figure 5.5 [110]. The reduction reactions where the metal cations M become zero-valent atoms involve the e^{-}_{aq} and H radical as well as the secondary radicals formed if a scavenger elements of OH$\cdot$ are added to the solution to prevent the oxidation reaction. Otherwise, it is the oxidation process which shall take place to oxidize the ions M or their atomic versions to a higher state of oxidation.

5.5.2. Physical Mechanism

In the discharge regime, the electrochemical system is generating a low temperature localized plasma around the electrode. In the formed gas film around the electrode, the thermal collisions of the atoms will ionize some of the present atoms in that gaseous media. Some of the electrons that are normally bound to the atoms will be ejected from the atoms and will form an electron gas that co-exists with the gas of atomic ions and neutral atoms. This state of matter is called plasma and more specifically for the electrochemical discharges, a partial thermodynamic equilibrium low temperature plasma [111, 114]. By analogy with the non-thermal plasma of electrical discharges, the electron temperature is of much higher magnitude than the temperature of the ions and the neutral gas formed around the cathodic polarized electrode. This temperature difference is explained by the small kinetic energy transfer due to the low collision frequency of the electrons emitted by the cathode and the ions present in the solution. Because the electronic temperature stays much higher than the temperature of the heavy ionic particles and gaseous molecules, which is equal or slightly above the room temperature, the phenomenon is a localized cold plasma.

In the volume around the electrode, the charged particles, i.e. electrons and ions, are in motion due to the applied potential difference between the two electrodes and they are subject to various elementary reactions such that electron-ion collisions, charge transfer, among others, due to ion-ion collisions and radiative and atomic processes. The rates of these processes depend on electron and atom energy distribution functions and on both electron and atom densities. Due to this energetic environment within the gas film and at its boundaries, the disintegration or erosion process of the electrode takes place generating clusters by cascade collisions of the ionic species at the upper atom planes of the electrode. Nanosuspensions of particles obtained by electrochemical discharges may be of material sputtering origin.

Another mechanism can be deducted from the electron microscopy observations of fine powder collected from the waste of electrical discharge machining (EDM) [115]. The powder contains spherical metal particles of size distribution ranging from micrometers down to the nanometre scale. By analogy, we suggest that during electrochemical discharges, localized melting of the electrode material followed by rapid cooling of the disintegrated parts in the electrolyte is also a possible route to build nanoparticles during the electrochemical discharge phenomenon. This mechanism is most likely dominant for high cell potentials and probably explains the observations of the spherical nanoballs done by Toriyabe et al. [111]. In addition, lattice defects in the nano-sized particles were observed by Toriyabe et al. and by the authors [109, 110] where for example dislocations defects are illustrated in figure 5.2(b).

Chapter 6

Conclusion

From conventional to high current density water electrolysis, the electrochemical discharge phenomenon exhibits remarkable characteristics with great potential for academic and industrial applications. The excess of heat production, the deviation from Faraday's law of the formation rates of chemical species, to the production of species novel for traditional electrolysis, the electric interruptions as well as the light emission all constitute the main features of this phenomenon.

A wide range of applications, from the macroscopic down to the nanoscopic scale, are developed by using and controlling those physicochemical properties of the electrochemical discharges. Promising techniques for large-scale works include micro-precision machining, nanostructure manufacture, surface treatment, and synthetic chemistry, to cite a few. The focus of this contribution is on the first two methods, as the authors are more familiar with the current views in those fields of research.

Comparing with the arc discharge technique under ultra-low pressure or the electronic irradiations process, the electrochemical discharge technique to synthesize nano-sized particles is simpler, cheaper and faster to operate. From the experimental and characterization sections, the electrochemical discharges appear to be a promising route in this field of science. As the aim of developing a new method for the fabrication of nanoparticles is to tailor the dimension of the obtained particles, the identification of the parameters controlling the process is fundamental. The authors consider the applied terminal voltage, the material electronic structure, the nature and the temperature of the electrolytic medium as major factors for the particles size control. Those parameters, intrinsic to the employed electrochemical cell, can be considered as second level contributors in the phenomenon since the stability and dynamics of the gas film are also governed by the control of the thermodynamic forces and fluxes related to the temperature, the chemical or electrochemical potential and to the electric environment of the system.

To conclude, we would like to emphasize that the cited bibliography contains numerous ideas and research directions, but we had to compromise by only selecting some of them, which are generally considered to be most important and relevant. In studying this contribution, the reader will realize that the authors have tried to combine together, in a simplified manner, both experimental and theoretical studies to bring a summary of the knowledge in this field as it stands today. Nevertheless, it has to be emphasized that there remains a large gap between the existing knowledge of this subject and the full potential of this complex electrochemical phenomenon.

Acknowledgement

The authors gratefully acknowledge the financial support from the Natural Sciences and Engineering Research Council of Canada.

References

[1] J.Z. Gao, X.Y. Wang, Z.A. Hu, J.G. Hou, and Q.F. Lu. A review on chemical effects in aqueous solution induced by plasma with glow discharge. *Plasma Science and Technology*, 3:765–774, 2001.

[2] A.L. Yerokhin, X. Nie, A. Leyland, A. Matthews, and S.J. Dowey. Plasma electrolysis for surface engineering. *Surface and Coatings Technology*, 122:73–93, 1999.

[3] R. Wüthrich. *Micromachining using electrochemical discharge phenomenon: fundamentals and applications of spark assisted chemical engraving*. Micro & nano technologies. USA: William Andrew, 2009.

[4] R. Wüthrich and V Fascio. Machining of non-conductive materials using electrochemical discharge phenomenon - an overview. *International Journal of Machine Tools and Manufacture*, 45:1095–1108, 2005.

[5] A. Lal, H. Bleuler, and R Wüthrich. Fabrication of metallic nanoparticles by electrochemical discharges. *Electrochemistry Communications*, 10: 488–491, 2008.

[6] R. Wüthrich and Ph. Mandin. Electrochemical discharges-discovery and early applications. *Electrochimica Acta*, 54:4031–4035, 2009.

[7] H. Fizeau and L. Foucault. Recherches sur l'intensité de la lumière émise par le charbon dans l'expérience de davy. *Comptes Rendus des Séances de l'Académie des Sciences*, 18:746–754, 1844.

[8] P. Hoho. Phénomène calorifique produit par le courant électrique au contact d'un solide et un liquide. *La Lumière Electrique*, 52(16):113–120, 1894.

[9] H. Bagard. A. Wehnelt. – Ein elektrolytischer stromunterbrecher (Interrupteur électrolytique) – Wied. Ann., LXVIII, p.233-272. *Journal de Physique Théorique Appliquée*, 8:438–444, 1899.

[10] A. Hickling and M.D. Ingram. Contact glow-discharge electrolysis. *Transactions of the Faraday Society*, 60:783–793, 1964.

[11] K. Harada, S. Igari, M. Takasaki, and A. Shimoyama. Reductive fixation of molecular nitrogen by glow discharge against water. *Journal of the Chemical Society, Chemical Communications*, 17:1384–1385, 1986.

[12] K. Harada and T. Iwasaki. Syntheses of amino acids from aliphatic carboxylic acid by glow discharge electrolysis. *Nature*, 250:426–428, 1974.

[13] K. Harada and S. Suzuki. Formation of amino acids from elemental carbon by contact glow discharge electrolysis. *Nature*, 266:275–276, 1977.

[14] Rolf Wüthrich, A. Lal, and H. Bleuler. A table-top set-up for nanoparticles production with electrochemical discharges. In *5th International Workshop on Microfactories*, 2006.

[15] H. Vogt. *Comprehensive Treatise of Electrochemistry*, volume 6, pages 445–489. New York, USA: Plenum, 1983.

[16] V. Fascio. *Etude de la microstructuration du verre par étincelage assistée par attaque chimique: une approche electrochimique*. PhD thesis, Swiss Federal Institute of Technology, 2002.

[17] R. Wüthrich, L.A. Hof, A. Lal, K. Fujisaki, H. Bleuler, Ph. Mandin, and G. Picard. Physical principles and miniaturization of spark assisted chemical engraving (sace). *Journal of Micromechanics and Microengineering*, 15:268–275, 2005.

[18] P. Boissonneau and P. Byrne. An experimental investigation of bubble-induced free convection in a small electrochemical cell. *Journal of Applied Electrochemistry*, 30:767–775, 2000.

[19] H. Vogt. The rate of gas evolution at electrodes—ii. an estimate of the efficiency of gas evolution on the basis of bubble growth data. *Electrochimica Acta*, 29:175–180, 1984.

[20] H. Vogt. The rate of gas evolution of electrodes—i. an estimate of the efficiency of gas evolution from the supersaturation of electrolyte adjacent to a gas-evolving electrode. *Electrochimica Acta*, 29:167–173, 1984.

[21] H. Vogt. The problem of the departure diameter of bubbles at gas-evolving electrodes. *Electrochimica Acta*, 34:1429–1432, 1989.

[22] L. J. J. Janssen. Behaviour of and mass transfer at gas-evolving electrodes. *Electrochemica Acta*, 34:161–169, 1989.

[23] J. Eigeldinger and H. Vogt. The bubble coverage of gas-evolving electrodes in a flowing electrolyte. *Electrochemica Acta*, 45:4449–4456, 2000.

[24] J Venczel. Über den gasblasen bei elektrochemischen prozessen. *Electrochimica Acta*, 15:1909–1920, 1970.

[25] H. Vogt. The incremental ohmic resistance caused by bubbles adhering to an electrode. *Journal of Applied Electrochemistry*, 13:87–88, 1983.

[26] R. Wüthrich. *Spark assisted chemical engraving-a stochastic modeling approach*. PhD thesis, Swiss Federal Institute of Technology, 2002.

[27] R. Wüthrich, Ch. Comninellis, and H. Bleuler. Bubble evolution on vertical electrodes under extreme current densities. *Electrochemica Acta*, 50 (5242-5246), 2005.

[28] R. Wüthrich and L.A. Hof. The gas film in spark assisted chemical engraving (SACE) – a key element for micro-machining applications. *International Journal of Machine Tools and Manufacture*, 46(7-8):828–835, 2006. ISSN 0890-6955.

[29] E. Klupathy. Zur theorie des wehneltunterbrechers. *Annalen der Physik*, 9:147, 1902.

[30] Ch. Guilpin. PhD thesis, Université Paris VI, 1975.

[31] D. Stauffer and A. Aharony. *Introduction to Percolation Theory*. London, UK, Taylor & Francis, 1998.

[32] I. Basak. *Electrochemical discharge machining mechanism and a scheme for enhancing material removal capacity*. PhD thesis, Indian Institute of Technology, 1991.

[33] H. Vogt and J. Thonstad. Review of the causes of anode effects in aluminium electrolysis. *Aluminium*, 79:98–102, 2003.

[34] H. Vogt. The anode effect as a fluid dynamic problem. *Journal of Applied Electrochemistry*, 29:137–145, 1999.

[35] H. H. Kellogg. Anode effect in aqueous electrolysis. *Journal of The Electrochemical Society*, 97(4):133–142, 1950. URL `http://link.aip.org/link/?JES/97/133/1`.

[36] Min-Seop Han, Byung-Kwon Min, and Sang Jo Lee. Geometric improvement of electrochemical discharge micro-drilling using an ultrasonic-vibrated electrolyte. *Journal of Micromechanics and Microengineering*, 19(6):065004, 2009. URL `http://stacks.iop.org/0960-1317/19/i=6/a=065004`.

[37] A. Allagui and R. Wüthrich. Gas film formation time and gas film life time during electrochemical discharge phenomenon. *Electrochimica Acta*, 54(23):5336–5343, 2009. ISSN 0013-4686.

[38] R. El-Haddad and R. Wüthrich. A mechanistic model of the gas film dynamics during the electrochemical discharge phenomenon. *Journal of Applied Electrochemistry*, 40:1853–1858, 2010. ISSN 0021-891X. URL `http://dx.doi.org/10.1007/s10800-010-0141-7`. 10.1007/s10800-010-0141-7.

[39] A. Hickling and M. D. Ingram. Glow-discharge electrolysis. *Journal of Electroanalytical Chemistry*, 8(1):65–81, 1964. ISSN 0022-0728.

[40] Y. S. Liao and W. Y. Peng. Study of hole-machining on pyrex wafer by electrochemical discharge machining (ecdm). *Materials Science Forum*, 505-507(2):1207–1212, 2006.

[41] George V. Buxton. *Radiation chemistry: from basics to applications in material and life sciences*, chapter 1. L'Actualité Chimique Livres. EDP Sciences, 2008.

[42] Christian Guilpin and Jocelyne Garbarz-Olivier. Analyse de la lumière émis aux électrodes pendant les effects d'électrode, dans des solutions aqueuses d'électrolyte. *Spectrochimica Acta Part B: Atomic Spectroscopy*, 32(3-4):155–164, 1977.

[43] Edwin J. Hart and J. W. Boag. Absorption spectrum of the hydrated electron in water and in aqueous solutions. *Journal of the American Chemical Society*, 84(21):4090–4095, 1962.

[44] David C. Walker. The hydrated electron. *Quarterly Reviews of the Chemical Society*, 21:79–108, 1967.

[45] J. Gubkin. Electrolysche metallabscheidung an der freien oberflache eine salzlösung. *Annalen der Physik*, 32(3):114–115, 1887.

[46] S. K. Sengupta, R. Singh, and A. K. Srivastava. A study on the origin of nonfaradaic behavior of anodic contact glow discharge electrolysis: The relationship between power dissipated in glow discharges and nonfaradaic yields. *Journal of the Electrochemical Society*, 145:2209–2213, 1998.

[47] M. D. Klapkiv. State of an electrolytic plasma in the process of synthesis of oxides based on aluminum. *Material Science*, 31:494–499, 1995.

[48] M. D. Klapkiv, H. M. Nykyforchyn, and V. M. Posuvailo. Spectral analysis of an electrolytic plasma in the process of synthesis of aluminum oxide. *Material Science*, 30:333–343, 1995.

[49] M. Tezuka and M. Iwasaki. Liquid-phase reactions induced by gaseous plasma. decomposition of benzoic acids in aqueous solution. *Plasmas & Ions*, 1(23-26), 1999.

[50] A. L. Yerokhin, L. O. Snizhko, N. L. Gurevina, A. Leyland, A. Pilkington, and A. Matthews. Discharge characterization in plasma electrolytic oxidation of aluminium. *Journal of Physics D: Applied Physics*, 36:2110–2120, 2003.

[51] Gao Jin-zhang, Hu Zhong-ai, Wang Xiao-yan, Hou Jing-guo, Lu Xiao-quan, and Kang Jing-wan. Degradation of α-naphthol by plasma in aqueous solution. *Plasma Science and Technology*, 3(1):641–646, 2001. URL `http://stacks.iop.org/1009-0630/3/641`.

[52] M. Tezuka and M. Iwasaki. Oxidative degradation of phenols by contact glow discharge electrolysis. *Electrochemistry*, 65:1057–1060, 1997.

[53] Meguru Tezuka and Masakazu Iwasaki. Plasma induced degradation of chlorophenols in an aqueous solution. *Thin Solid Films*, 316(1-2):123 – 127, 1998. ISSN 0040-6090.

[54] Jinzhang Gao, Xiaoyan Wang, Zhongai Hu, Hualing Deng, Jingguo Hou, Xiaoquan Lu, and Jingwan Kang. Plasma degradation of dyes in water with contact glow discharge electrolysis. *Water Research*, 37(2):267 – 272, 2003. ISSN 0043-1354.

[55] W. R. Grove. On the electro-chemical polarity of gases. *Philosophical Transactions of the Royal Society of London*, 142:87–101, 1852. ISSN 02610523. URL `http://www.jstor.org/stable/108536`.

[56] K. H. Kingdon and Irving Langmuir. The removal of thorium from the surface of a thoriated tungsten filament by positive ion bombardment. *Physical Review*, 22(2):148–160, 1923.

[57] M. W. Thompson. The contribution of collision cascades to sputtering and radiation damage. *Philosophical Transactions: Mathematical, Physical and Engineering Sciences*, 362(1814):5–28, 2004. ISSN 1364503X. URL `http://www.jstor.org/stable/4142334`.

[58] Irving Langmuir. The constitution and fundamental properties of solids and liquids. part i. solids. *Journal of the American Chemical Society*, 38 (11):2221–2295, 1916.

[59] H. Massey and E. Burhop. *Electronic and ionic impact phenomena.* Clarendon Press, Oxford, 1952.

[60] L. R. Ingersoll and L. O. Sordahl. Temperature of the cathode as a factor in the sputtering process. *Physical Review*, 32(4):649–656, 1928.

[61] Gottfried K. Wehner. Momentum transfer in sputtering by ion bombardment. *Journal of Applied Physics*, 25(2):270–271, 1954. URL `http://link.aip.org/link/?JAP/25/270/1`.

[62] Erich B. Henschke. New collision theory of cathode sputtering of metals at low ion energies. *Physical Review*, 106(4):737–753, 1957.

[63] R. H. Silsbee. Focusing in collision problems in solids. *Journal of Applied Physics*, 28(11):1246–1250, 1957. URL http://link.aip.org/link/?JAP/28/1246/1.

[64] Raúl A. Baragiola. Sputtering: survey of observations and derived principles. *Philosophical Transactions: Mathematical, Physical and Engineering Sciences*, 362(1814):29–53, 2004. ISSN 1364503X. URL http://www.jstor.org/stable/4142335.

[65] G. Betz and W. Husinsky. Modelling of cluster emission from metal surfaces under ion impact. *Philosophical Transactions: Mathematical, Physical and Engineering Sciences*, 362(1814):177–194, 2004. ISSN 1364503X. URL http://www.jstor.org/stable/4142343.

[66] John S. Colligon. Ion-assisted sputter deposition. *Philosophical Transactions: Mathematical, Physical and Engineering Sciences*, 362(1814): 103–116, 2004. ISSN 1364503X. URL http://www.jstor.org/stable/4142338.

[67] S. Facsko, T. Dekorsy, C. Koerdt, C. Trappe, H. Kurz, A. Vogt, and H. L. Hartnagel. Formation of ordered nanoscale semiconductor dots by ion sputtering. *Science*, 285(5433):1551–1553, 1999. ISSN 00368075. URL http://www.jstor.org/stable/2898112.

[68] G. Falcone. Ejection process in collisional sputtering. *Physical Review B*, 33(7):5054–5056, 1986.

[69] Don E. Harrison. Theory of the sputtering process. *Physical Review*, 102 (6):1473–1480, 1956.

[70] Peter Sigmund. Theory of sputtering. I. sputtering yield of amorphous and polycrystalline targets. *Physical Review*, 184(2):383–416, 1969.

[71] Kenkichiro Kobayashi, Yasumasa Tomita, and Masahiro Sanmyo. Electrochemical generation of hot plasma by pulsed discharge in an electrolyte. *The Journal of Physical Chemistry B*, 104(26):6318–6326, 2000.

[72] M. Mousa, A. Allagui, H. D. Ng, and R. Wuthrich. The effect of thermal conductivity of the tool electrode in spark-assisted chemical engraving gravity-feed micro-drilling. *Journal of Micromechanics and Microengineering*, 19(1):015010, 2009.

[73] H. Kurafuji and H. Suda. Electrical discharge drilling of glass. *Annals of the CIRP*, 16:415–419, 1968.

[74] K. Allesu, A. Ghosh, and M. K. Muju. A preliminary qualitative approach of a proposed mechanism of material removal in electrical machining of glass. *European journal of mechanical engineering*, 36(3): 201–207, 1991.

[75] S Tandon, V K Jain, P Kumar, and K P Rajurkar. Investigations into machining of composites investigations into machining of composites investigations into machining of composites. *Precision Engineering*, 12(4): 227–238, 1990.

[76] H. Langen, V. Fascio, R. Wüthrich, and D. Viquerat. Three-dimensional structuring of pyrex glass devices-trajectory control. In *International Conference of the European Society for Precision Engineering and Nanotechnology (EUSPEN)*, pages 435–438, 2002.

[77] B. Bhattacharyya, B.N. Doloi, and S.K. Sorkhel. Experimental investigations into electrochemical discharge machining (ecdm) of nonconductive ceramic materials. *Journal of Materials Processing Technology*, 24:145, 1999.

[78] N.H. Cook, G.B. Foote, P. Jordan, and B.N. Kalyani. Experimental studies in electro-machining. *Trans. ASME, Journal of Engineering for Industry*, pages 945–950, 1973.

[79] V. K. Jain and S. K. Chak. Electrochemical spark trepanning of alumina and quartz. *Machining Science and Technology: An International Journal*, 4(2):277–290, 2000.

[80] Hyung J. Lim, Young M. Lim, Soo Hyun Kim, and Yoon Keun Kwak. Self-aligned microtool and electrochemical discharge machining (ecdm) for ceramic materials. volume 4416, pages 348–353, Yokohama, Japan, 05 2001. SPIE.

[81] Hitoshi Tokura, Iwao Kondoh, and Masanori Yoshikswa. Ceramic material processing by electrical discharge in electrolyte. *Journal of Materials Science*, 24(3):991–998, 03 1989. URL `http://dx.doi.org/10.1007/BF01148788`.

[82] V. Fascio, R. Wüthrich, D. Viquerat, and H. Langen. 3d microstructuring of glass using electrochemical discharge machining (ecdm). In *International Symposium on Micromechatronics and Human Science*, pages 179–183, 1999.

[83] V.K. Jain, P.S. Rao, S.K. Choudhury, and K.P. Rajurkar. Experimental investigations into travelling wire electrochemical spark machining (tw-ecsm) of composites. *Trans. ASME, Journal of Engineering for Industry*, 113(1):75–84, 1991.

[84] M. Jalali, P. Maillard, S. P. Mishra, L. Kadem, and R. Wüthrich. Modeling gravity feed micro hole drilling by electrochemical discharge. In *2nd International Conference on Micro-Manufacturing*, Greenville, South Carolina, USA., September 10-13 2007.

[85] M Jalali, P Maillard, and R Wüthrich. Toward a better understanding of glass gravity-feed micro-hole drilling with electrochemical discharges. *Journal of Micromechanics and Microengineering*, 19(4):045001, 2009. URL `http://stacks.iop.org/0960-1317/19/i=4/a=045001`.

[86] Rolf Wüthrich, Ulrich Spaelter, Y Wu, and Hannes Bleuler. A systematic characterization method for gravity-feed micro-hole drilling in glass with spark assisted chemical engraving (SACE). *Journal of Micromechanics and Microengineering*, 16(9):1891–1896, 2006.

[87] C.T. Yang, S.S. Ho, and B.H. Yan. Micro hole machining of borosilicate glass through electrochemical discharge machining (ecdm). *Key Engineering Materials*, 196:149–166, 2001.

[88] V. Fascio, R. Wüthrich, and H. Bleuler. Spark assisted chemical engraving in the light of electrochemistry. *Electrochemica Acta*, 49:3997–4003, 2004.

[89] H. Tokura, I. Kondoh, and M. Yoshikswa. Ceramic material processing by electrical discharge in electrolyte. *Journal of Materials Science*, 24: 991–998, March 1989.

[90] V.K. Jain and S. Adhikary. On the mechanism of material removal in electrochemical spark machining of quartz under different polarity conditions. *Journal of Materials Processing Technology*, 200:460–470, 2008.

[91] T.F. Didar, A. Dolatabadi, and R. Wüthrich. Local hardness and density variation in glass substrates machined with spark assisted chemical engraving (sace). *Materials Letters*, 63:51–53, 2009.

[92] J. West and A. Jadhav. Ecdm methods for fluidic interfacing through thin glass substrates and the formation of spherical microcavities. *Journal of Micromechanics and Microengineering*, 17:403–409, 2007.

[93] A. Morrison, L. Rodrigues, and R. Wüthrich. Reducing variability in spark assisted chemical engraving gravity feed drilling of glass. In *Microsystems and Nanoelectronics Research Conference, 2008. MNRC 2008. 1st*, pages 161–164, 2008.

[94] P. Maillard, B. Despont, H. Bleuler, and R. Wüthrich. Geometrical characterization of microholes drilled in glass by gravity-feed with spark assisted chemical engraving (sace). *Journal of Micromechanics and Microengineering*, 17:1343–1349, 2007.

[95] Min-Seop Han, Byung-Kwon Min, and Sang Jo Lee. Improvement of surface integrity of electro-chemical discharge machining process using powder-mixed electrolyte. *Journal of Materials Processing Technology*, 191(1-3):224–227, 8 2007. URL `http://www.sciencedirect.com/science/article/B6TGJ-4N61FST-1/2/b14aff94a13c49a3b06952753d9a7cc5`.

[96] Eunice S Lee, Dwight Howard, Enzhu Liang, Scott D Collins, and Rosemary L Smith. Removable tubing interconnects for glass-based microfluidic systems made using ecdm. *Journal of Micromechanics and Microengineering*, 14(4):535–541, 2004. URL `http://stacks.iop.org/0960-1317/14/535`.

[97] Tohid Fatanat Didar, Ali Dolatabadi, and Rolf Wuthrich. Characterization and modeling of 2d-glass micro-machining by spark-assisted chemical engraving (sace) with constant velocity. *Journal of Micromechanics and Microengineering*, 18(6):065016, 2008.

[98] R. Wuthrich, B. Despont, P. Maillard, and H. Bleuler. Improving the material removal rate in spark-assisted chemical engraving (sace) gravity-feed micro-hole drilling by tool vibration. *Journal of Micromechanics and Microengineering*, 16(11):N28–N31, 2006.

[99] R. Wüthrich, V. Fascio, D. Viquerat, and H. Langen. In situ measurement and micromachining of glass. In *Micromechatronics and Human Science, 1999. MHS '99. Proceedings of 1999 International Symposium*, pages 185–191, 1999.

[100] Naveen Gautam and Vijay K. Jain. Experimental investigations into ecsd process using various tool kinematics. *International Journal of Machine Tools and Manufacture*, 38(1-2):15–27, 1998/2// . URL http://www.sciencedirect.com/science/article/B6V4B-3SYRYYM-2/2/228b4c079df3e97bf38fa001c2fc04ff.

[101] Zhi-Ping Zheng, Hsin-Chuan Su, Fuang-Yuan Huang, and Biing-Hwa Yan. The tool geometrical shape and pulse-off time of pulse voltage effects in a pyrex glass electrochemical discharge microdrilling process. *Journal of Micromechanics and Microengineering*, 17(2):265–272, 2007.

[102] Xuan Doan Cao, Bo Hyun Kim, and Chong Nam Chu. Micro-structuring of glass with features less than 100 μm by electrochemical discharge machining. *Precision Engineering*, 33(4):459 – 465, 2009. ISSN 0141-6359.

[103] C. T. Yang, S. L. Song, B. H. Yan, and F. Y. Huang. Improving machining performance of wire electrochemical discharge machining by adding sic abrasive to electrolyte. *International Journal of Machine Tools and Manufacture*, 46(15):2044–2050, 12 2006. URL http://www.sciencedirect.com/science/article/B6V4B-4JHMFGF-1/2/6d84b04769e3c8f5c940cf34aa1fe2c7.

[104] Dae-Jin Kim, Yoomin Ahn, Seoung-Hwan Lee, and Yong-Kweon Kim. Voltage pulse frequency and duty ratio effects in an electrochemical discharge microdrilling process of pyrex glass. *International Journal of Machine Tools and Manufacture*, 46(10):1064–1067, 8 2006. URL http://www.sciencedirect.com/science/article/B6V4B-4H6XKX9-5/2/25dae1f5d2ff1f9719005eaeeb00ede0.

[105] Zhi-Ping Zheng, Wei-Hsin Cheng, Fuang-Yuan Huang, and Biing-Hwa Yan. 3d microstructuring of pyrex glass using the electrochemical discharge machining process. *Journal of Micromechanics and Microengineering*, 17(5):960–966, 2007.

[106] Min-Seop Han, Byung-Kwon Min, and Sang Jo Lee. Modeling gas film formation in electrochemical discharge machining processes using a side-insulated electrode. *Journal of Micromechanics and Microengineering*, 18(4), 2008.

[107] H. Kawamura, K. Moritani, and Y. lto. Discharge electrolysis in molten chloride: formation of fine silver particles. *Plasmas & Ions*, 1(1):29–36, 1998. ISSN 1288-3255. URL `http://www.sciencedirect.com/science/article/B6VPX-3W8RYJ3-4/2/24ff9bd64f0784533826fdee28bf4da3`.

[108] T. Oishi, H. Kawamura, and Y. Ito. Formation and size control of titanium particles by cathode discharge electrolysis of molten chloride. *Journal of Applied Electrochemistry*, 32(7):819–824, 2002.

[109] A. Allagui, R. Wüthrich, and E.A. Baranova. Copper and nickel nanoparticles: synthesis by electrochemical discharges. In *Microsystems and Nanoelectronics Research Conference, 2008. MNRC 2008. 1st*, pages 65–68, 2008.

[110] Rolf Wüthrich and Anis Allagui. Building micro and nanosystems with electrochemical discharges. *Electrochimica Acta*, 55(27):8189–8196, 2010.

[111] Y. Toriyabe, S. Watanabe, S. Yatsu, T. Shibayama, and T. Mizuno. Controlled formation of metallic nanoballs during plasma electrolysis. *Applied Physics Letters*, 91(4):041501–041503, 2007.

[112] M Kelm, J Lilie, A Henglein, and E Janata. Pulse radiolytic study of nickel(1+) ion. nickel-carbon bond formation. *The Journal of Physical Chemistry*, 78:882–887, 1974.

[113] Jacqueline Belloni and Hynd Remita. *Radiation chemistry: from basics to applications in material and life sciences*, chapter 7. L'Actualité Chimique Livres. EDP Sciences, 2008.

[114] J. Meichsner. *Plasma physics, confinement, transport and collective effects*, volume 670 of *Lecture Notes in Physics*, chapter 5. Springer, 2005.

[115] M. Jalali, A.G. Terpstra, and R. Wüthrich. Fabrication of metallic nanoparticles through waste of micro-edm. In *9th Euspen International Conference*, 2009.

Index

T

U

V

W

X

Y